Solid State Composites and Hybrid Systems

Fundamentals and Applications

Solid State Composites and Hybrid Systems

Fundamentals and Applications

Edited by
Rada Savkina
Larysa Khomenkova

CRC Press
Taylor & Francis Group
Boca Raton London New York

CRC Press is an imprint of the
Taylor & Francis Group, an **informa** business

CRC Press
Taylor & Francis Group
6000 Broken Sound Parkway NW, Suite 300
Boca Raton, FL 33487-2742

First issued in paperback 2020

© 2019 by Taylor & Francis Group, LLC
CRC Press is an imprint of Taylor & Francis Group, an Informa business

No claim to original U.S. Government works

ISBN 13: 978-0-367-57116-0 (pbk)
ISBN 13: 978-0-8153-8603-2 (hbk)

Library of Congress Cataloging-in-Publication Data

Names: Savkina, Rada, editor. | Khomenkova, Larysa, editor.
Title: Solid state composites and hybrid systems : fundamentals and applications / edited by Rada Savkina, Larysa Khomenkova.
Description: Boca Raton, FL : CRC Press, Taylor & Francis Group, [2018] | Includes bibliographical references and index.
Identifiers: LCCN 2018011865| ISBN 9780815386032 (hardback ; alk. paper) | ISBN 0815386036 (hardback ; alk. paper) | ISBN 9781351176071 (eBook) | ISBN 1351176072 (eBook) | ISBN 9781351176064 (eBook Adobe Reader) | ISBN 1351176064 (eBook Adobe Reader) | ISBN 9781351176057 (eBook ePub) | ISBN 1351176056 (eBook ePub) | ISBN 9781351176040 (eBook Mobipocket) | ISBN 1351176048 (eBook Mobipocket)
Subjects: LCSH: Composite materials. | Polymeric composites. | Nanocomposites (Materials)
Classification: LCC TA418.9.C6 S63 2018 | DDC 620.1/18--dc23
LC record available at https://lccn.loc.gov/2018011865

Visit the Taylor & Francis Web site at
http://www.taylorandfrancis.com

and the CRC Press Web site at
http://www.crcpress.com

To our mothers

Contents

Preface

SOLID STATE composite and hybrid structures are currently receiving significant interest and becoming one of the most attractive topics in the interdisciplinary fields of nanoscience and nanotechnology, surface and material science, bioscience and bioengineering, etc. Energy harvesting and storage, the environment and human health have emerged as strategic priorities, not only in research, but in all aspects of our lives. Multifunctional composite/hybrid materials have already demonstrated promise in offering solutions in each of these priority areas.

The progress in materials science and technology has come to a stage where macroscopic properties of the materials are determined by their micro- and nanoscopic nature. The development of materials from their fundamental understanding to manufacturing affects not only the community of researchers and the way they do research but also ordinary people and their daily life, for instance, via cell phones, solar cell-powered gadgets, information storage, nanomedicine, pain-free operations, controlled drug delivery, etc.

Composites and hybrid systems, combined from inorganic and/or organic constituents in different ways, are gaining increasing attention because of their intriguing fundamental physics at the nanoscale level. That's why the search for compositional materials with novel functionality is one of the most challenging projects. The creation of such materials is now possible with a high degree of complexity because of impressive developments in these technologies. As a result, new compositional material (so-called "composites") will have characteristics which unexpectedly differ from parent constituents. It will allow the integration of several key functions in a single structure and more complex properties for the new devices.

This book presents for the reader the review of the original works and the patents received over the last several years. It covers a wide range of materials, techniques and approaches that could be interesting both for experienced researchers and for young scientists.

This book consists of six chapters, each written as an independent part with an introduction, a main body of text, conclusions and a list of references, and each can be read separately. Chapter 1 concerns general consideration of "composites" and "hybrid systems", the methods for their fabrication and characterisation, and some important fields of the application. Chapter 2 addresses the zirconia-based nanocomposites that offer different applications, such as solid electrolytes, gas sensors, catalysts, biological labelling, fungicides, optical communication, light-emitting devices, etc. Most of these applications require other oxides (such as Y_2O_3, CuO, Sc_2O_3, NiO, CeO_2, etc.) to be added to ZrO_2. Thus, the effect of doping and co-doping of ZrO_2 with different aliovalent impurities on structural, optical and chemical properties of the final nanocomposite is described and possible applications are demonstrated. Chapter 3 demonstrates the main advantages and the fields of applications of the nanocomposites based on the semiconductor quantum dots (CdSe, $AgInS_2$, $CuInS_2$, core-shell CdSe/ZnS, $AgInS_2$/ZnS and $CuInS_2$/ZnS QDs) embedded in biocompatible polymer hosts (gelatine and polyvinyl alcohol [PVA]). The main results obtained by the author of the chapter are reviewed and demonstrate that better passivation of the QD surface states is achieved with gelatine functional groups allowed enhancement of the luminescence intensity and photo-stability of the QD-gelatine system as compared to those of QD-PVA. The utility of such nanocomposites in bio-imaging, sensing and light-emitting devices, as well as their toxicity investigations, are discussed. In Chapter 4, the latest advances in the field of ultrasound-assisted preparation of the solid state composites and hybrid systems are discussed. This chapter answers the question: "What is cavitation, and how does this phenomenon work in the area of the materials manufacturing?" The recent achievements in this field in the sonochemical synthesis of the composite/hybrid materials are shown. The advantages of this method include the low reaction temperatures, faster polymerisation rates and higher molecular weight of polymers which are used for production of the inorganic-organic composites with high complexity. Also discussed is the method of the high-frequency ultrasound-assisted formation of the nanostructured $Si/SiO_2/(CaSiO_3)$ composite demonstrating strong optical emission in the visible spectral range and biocompatibility confirmed by hydroxyapatite formation after storage in the simulated body fluid solution. In Chapter 5, the review of the applied research dealing with ion implantation as a tool for modifying and enhancing the materials properties is presented. The focus is on a specific class of nanomaterials as metal nanoclusters embedded in semiconductor matrices (silicon, II-VI and III-V compounds) aiming at their application in photonics, magnetism, sensors, radiation detectors and infrared and sub-THz optoelectronics. Chapter 6 considers the composite materials based on the silicon and/or germanium nanocrystals embedded in different dielectrics (SiO_2, Al_2O_3, HfO_2). The fabrication of the nanocomposite thin films and superlattices by magnetron sputtering is addressed. The effect of fabrication conditions and annealing treatment on the structural, optical and luminescent properties of such materials and superlattices is considered. The mechanisms of the nanocrystals' formation via phase separation process are proposed. The possible application of these materials and the ways of monitoring their properties are discussed.

We hope to present the reader with a recent survey of the topic of composite materials and hybrid systems, as well as to depict the potential and broadness of the applications of such materials.

We would like to express our gratitude to all contributors for their willingness to share their experience and in-depth insights into this field, and for taking time out of their schedules to write excellent chapters. We would like to thank all our current and former colleagues who helped and advised us in conducting research described in this book. Also, we are thankful to the staff of Taylor & Francis/CRC Press, who invited us to write this book and helped immensely in its preparation for publication. And last, but not least, we are deeply grateful to our families for their support, patience and understanding over the many busy years of our scientific careers.

Rada Savkina and Larysa Khomenkova
Kyiv, Ukraine

Acknowledgements

The results that the authors have described in this book were obtained in different years under national and international projects. We would like to ackowledge the National Academy of Sciences of Ukraine (NASU), the Ministry of Education and Science of Ukraine, the French National Research Agency, the National Research Centre of France, the French–Ukraine bilateral program 'Dnipro' and the EU Seventh Framework Programme for their partial support of our research in:

- the project "Affine sensor system based on 'smart' hybrid nanocomplexes for determination of specific nucleic acid sequences" in the framework of target programme of the NASU, "'Intelligent' sensor devices of a new generation based on modern materials and technologies" (2018–2022, Ukraine);

- the project "Effect of the doping on structural, optical and electron-phonon properties and stability of anisotropic crystals" (No: 89452, 2018–2020, Ukraine);

- the project "Development of technology and equipment for the creation of infrared, THz and sub-THz radiation based on low-dimensional structures" in the framework of the programme "Search and creation of promising semiconductor materials and functional structures for nano- and optoelectronics" (No: III-41-17, 2017–2022, Ukraine);

- the project "Creation of bioactive composite structures on the basis of elementary (Si) and binary (A_3B_5) semiconductors with photovoltaic properties" in the framework of the programme "Search and creation of promising semiconductor materials and functional structures for nano- and optoelectronics" (No: III-41-17, 2017–2022, Ukraine);

- the project "Physics and technology of multifunctional materials and structures based on oxides of metals, silicon, III-V and II-VI compounds for application in modern devices of optoelectronics, microelectronics and UHF technology" (No. III-4-16; 2016–2020, Ukraine);

- the projects in the framework of the programme "Fundamental problems of new nanomaterials and nanotechnologies creation" (No: 11/16-H; 11/17-H; 11/18-H; 2015–2019, Ukraine);

- the project "Study and nanoanalysis of irradiation effects group" within the "Investments for the Future Programme" and the "GENESIS EQUIPEX Program*me*" (PIA, ANR, (ANR-11-EQPX-0020) and Normandy Region) (2014–2018, France);

- the project "High-k oxides doped with Si or Ge nanoclusters for microelectronic and photonic applications" in the framework of the Ukrainian–French bilateral programme "Dnipro", 2015–2016 edition (No. M/115-2015 and M/27-2016 (Ukraine), No. 34820QH (France));

- the project "High-k dielectric structures with modulated architecture for non-volatile memory application" in the framework of the *Programme Investissements d'Avenir* (ANR-11-IDEX-0002-02; ANR-10-LABX-0037-NEXT (2015, France);

- the European Union Seventh Framework Programme under Grant Agreement 312483 – ESTEEM2 (Integrated Infrastructure Initiative–I3) (2015, France);

- the project "Improvement of the methods for producing luminescent quantum-confined structures based on II-VI compounds for application in optoelectronics and medicine" in the framework of the programme "Development of technological methods for creation of new functional materials and structures for modern electronics, information technology and sensors" (No: III-41-12, 2012-2015, Ukraine);

- the project "Photovoltaic, optical and fluctuation phenomena in light-emitting wide bandgap semiconductor compounds, semiconductor structures of micro and nanoelectronics and the development of technological methods for their production" (No. III-4-11; 2011-2015, Ukraine).

Editors

Dr Rada Savkina is a senior staff researcher of the V. Lashkaryov Institute of Semiconductor Physics at the National Academy of Sciences of Ukraine (Kyiv, Ukraine). She earned an M.Sc. in solid state physics in 1993 from T. Shevchenko Kyiv University and a Ph.D. in solid state physics in 2002 from V. Lashkaryov Institute of Semiconductor Physics, where she was named senior researcher in solid state physics in 2013. She has valuable experience in developing new ideas from a fundamental understanding to the manufacturing of multifunctional composite materials and devices on their basis. She is an expert in the fabrication of materials by physical methods (laser deposition, ion beam processing and ultrasonic approach) as well as their characterisation with optical and electrical methods. Her research and development in the field of infrared detectors and vision systems are protected by five patents. She has more than 60 papers in refereed journals and two book chapters. She is a member of the editorial board of the international research journal *Recent Advances in Electrical & Electronic Engineering*. Her activity also concerns the elaboration of the concept of a new biocompatible photovoltaic cell by a sonochemical method.

Dr Larysa Khomenkova is a senior staff researcher of the V. Lashkaryov Institute of Semiconductor Physics at the National Academy of Sciences of Ukraine (Kyiv, Ukraine), as well as an associate professor of the Faculty of Natural Sciences at the National University of Kyiv–Mohyla Academy. She earned an M.Sc. in technology and materials in microelectronics in 1992 from T. Shevchenko Kyiv University and a Ph.D. in solid state physics in 1999 from V. Lashkaryov Institute of Semiconductor Physics, where she was named senior researcher in solid state physics in 2003. Her work is centred around the multifunctional composite materials, in particular their fabrication with physical methods (solid state reaction; physical vapour deposition; magnetron sputtering) and the materials' characterisation with optical, luminescent and electrical methods. Her interest also focuses on the elaboration of novel technological approaches for the creation of semiconductor and dielectric structures aiming at their photonic and microelectronic applications, as well as on the instability and degradation phenomena in the devices. Her activity also concerns the elaboration of the materials for alternative energy sources such as solid oxide fuel cells. She has more than 130 research papers and two book chapters. She is the editor of two books.

Contributors

Lyudmyla Borkovska is head of the laboratory of multifunctional composite materials in the Department of Semiconductor Heterostructures of the V. Lashkaryov Institute of Semiconductor Physics NASU, Kyiv, Ukraine.

Nadiia Korsunska is a professor in the laboratory of multifunctional composite materials in the Department of Semiconductor Heterostructures of the V. Lashkaryov Institute of Semiconductor Physics NASU, Kyiv, Ukraine.

Aleksej Smirnov is a senior researcher in the Department of Physics and Technology of Low-Dimensional Systems of the V. Lashkaryov Institute of Semiconductor Physics NASU, Kyiv, Ukraine.

Abbreviations

2D	two-dimensional
3D	three-dimensional
AES	Auger electron spectroscopy
AFM	atomic force microscopy
ATR	attenuated total reflection
BSA	bovine serum albumin
CCT	correlated colour temperature
CIE	Commission Internationale de l'Eclairage
CL	cathodoluminescence
CNTs	carbon nanotubes
CMOS	complementary metal–oxide–semiconductor
CRI	colour rendering index
CS	calcium silicate
DNA	deoxyribonucleic acid
EBL	electron-beam lithography
EF-TEM	energy-filtered transmission electron microscopy
EUV	extreme ultraviolet
FESEM	field emission scanning electron microscopy
FETs	field-effect transistors
FLA	flash lamp annealing
FTIR	Fourier transform infrared
GI XRD	grazing-incidence XRD
HA	hydroxyapatite
HIFU	High Intensity Focused Ultrasound
hrGO	highly reduced graphene oxide
HR-RSM	high-resolution reciprocal space mapping
HR-TEM	high-resolution transmission electron microscopy
IR	infrared
ITRS	International Technology for Semiconductors
LCD	liquid crystal display
LDH	layered double hydroxide
LED	light-emitting diode
LN2, LN$_2$	liquid nitrogen

LSC	luminescent solar concentrator
LWIR	long-wave infrared
MCT	HgCdTe
MOFs	metal-organic frameworks
MOSFET	metal-oxide-semiconductor field-effect transistor
MUA	mercaptoundecanoic acid
MWCNTs	multiwalled carbon nanotubes
MWIR	mid-wave infrared
NI	normal-incidence
NP	nanoparticle
NC	nanocrystal
NW	nanowire
NWIR	near-wave infrared
OI	oblique-incidence
PL	photoluminescence
PLE	photoluminescence excitation
PSS	sodium polystyrene sulphonate
PV	photovoltaic
PVD	physical vapour deposition
PZT	lead zirconate titanate, $Pb[Zr_xTi_{1-x}]O_3$ $(0 \leq x \leq 1)$
QD	quantum dots
QY	quantum yield
RC	rocking curves
RF	radio frequency
RMS	root-mean-square roughness
SBF	simulated body fluid
SCS	sonochemical synthesis
ScSZ	Scandia-stabilised zirconia
SEM	scanning electron microscopy
STEM	scanning transmission electron microscopy
THz	terahertz radiation
TOPO	trioctylphosphine oxide
ULSI	ultra large scale integration
US	ultrasound
USP	ultrasonic spray pyrolysis
UV	ultraviolet
XPS	X-ray photoelectron spectroscopy
XRD	X-ray diffraction
YSZ	yttria-stabilised zirconia

CHAPTER 1

Solid State Nanocomposites and Hybrid Systems

General Remarks

Larysa Khomenkova and Rada Savkina

CONTENTS

1.1 CLASSIFICATION

When two or more constituent materials are combined in one, a final *compositional material* can be built. Nowadays, the most common name for such materials is *composites*.

The individual components of the composite remain without their dissolving or blending into each other and can be distinct within the final structure. As for the constituent materials, they can be classified into two main types: *matrix* (or binder) and *reinforcement*. To build a composite, at least one portion of each type is required. The matrix material surrounds and supports the reinforcement materials by maintaining their relative positions. The reinforcements impart their special mechanical, physical and optical properties to enhance the matrix properties.

The compositional material has very different physical and chemical characteristics (Figure 1.1). It can be stronger, lighter or less expensive when compared to traditional materials. When aiming for a particular application of the composites, an appropriate combination of matrix and reinforcement materials should be chosen.

Based on the matrix material, composites can be categorised into polymer composites, ceramic composites, carbon composites and metal composites. Composites, consisting of organic–organic, organic–inorganic and inorganic–inorganic building blocks, can combine the properties from the parent constituents and generate new properties to fit the requirements of their functionality. Recent developments in inorganic and organic nanoparticle synthesis provide an abundance of building blocks that offers the construction of solid state composites and hybrid systems with unlimited possibilities (Kao et al. 2013; Heinz et al. 2017; Otero et al. 2017).

The properties of composite materials depend not only on those of individual building blocks, but also on their spatial organisation at different length scales. The mixing of the constituents can occur at macroscopic and microscopic levels. One of the examples

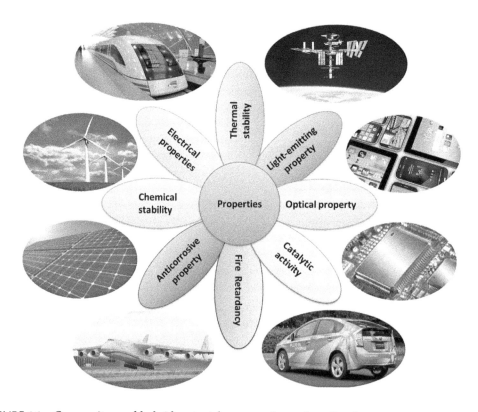

FIGURE 1.1 Composites and hybrid materials: properties and applications.

of macroscopic composites is large-scale glass-fibre–reinforced aluminium. The elaboration of such materials stimulated, for instance, the development of modern aircraft. This new composite is 25% stronger and 20% lighter than conventional airframe aluminium, so it offers the possibility of making lighter cars and aircraft, followed by less fuel use. One more example is the utilisation of carbon fibre composite nanomaterials within the composite laminate to enhance the radio frequency (RF) reflective properties – particularly at frequencies >20 GHz (www.eclipsecomposites.com). Using the laminate recipes, the increased stiffness of carbon fibre materials has enabled weight reduction of up to 70% and better mechanical strength (George et al. 2014) that facilitates greater precision on the parabolic surface when compared to aluminium (Figure 1.2). Carbon fibre materials have a much lower coefficient of thermal expansion that reduces parabolic surface distortion due to thermal gradients across the reflector surface.

The mixing at the microscopic scale leads to a more homogeneous material. This latter shows either characteristics in-between the two original phases or even new properties. For instance, the composites based on doped oxides offer advanced water purification (Heinz et al. 2017); the glasses reinforced with nanoparticles used for the smart windows (see Figure 1.3) provide significant energy savings and human health protection (Granqvist et al. 2014).

The mixing of different constituents as multilayered structures allowed the achievement of novel mechanical properties of the composite. For instance, a metal-graphene nanolayered composite utilises the key advantages of graphene, including ultra-high strength, modulus and 2D geometry as a strength enhancer (Kim et al. 2013). The mechanical properties of the synthesised metal-graphene nanolayered composites are measured using nanopillar compression testing.

The development of the microelectronics and photonics was stimulated by the elaboration of the solid state composites in which the mixing of the constituents occurs both at the microscopic and nanometre levels. For instance, the progress in the composite optical fibres achieved the high-speed data transfer and high-quality communication. As for microelectronics, its demand in aggressive miniaturisation of the devices forces the elaboration of different nanocomposites.

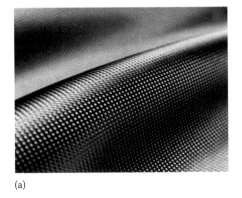

(a)

www.eclipsecomposites.com

(b)

FIGURE 1.2   Carbon fibre composite laminate (a) and parabolic antennas covered with such laminates (b). (From www.eclipsecomposites.com.)

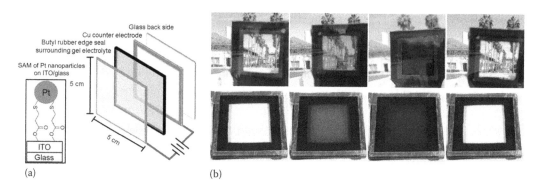

FIGURE 1.3   (a) Schematic of a 25 cm² two-electrode dynamic window containing a Cu-Ag gel electrolyte with an ITO-on-glass working electrode modified with an SAM of Pt nanoparticles (inset). (b) Photographs of the dynamic window during switching outside (top) and inside with a white background (bottom) after 0 s, 15 s and 30 s of metal electrodeposition at −0.6 V and subsequently 30 s of metal stripping at +0.8 V (from left to right). (Reproduced with permission from *Joule*, 1, Barile, C. J., et al., Dynamic windows with neutral colour, high contrast, and excellent durability using reversible metal electrodeposition, 133–145, Copyright 2017, Elsevier.)

This allows the existence of digital devices, small in physical size but able to store huge amounts of information; in the development of diagnostic devices and surgery instruments; the development of new energetic concepts, etc. Among these latter, one can address dynamic windows (Figure 1.3). They are able to switch between transparent and opaque states under applied bias. This phenomenon has applications in buildings, automobiles and transition lenses.

Figure 1.3 shows dynamic windows based on the reversible electrodeposition of Cu and a second metal on transparent indium tin oxide electrodes modified by Pt nanoparticles. Three-electrode cyclic voltammetry experiments reveal that the system possesses high Coulombic efficiency (99.9%), indicating that the metal electrodeposition and stripping processes are reversible.

These devices switch between transparent and opaque states within 3 minutes and are able to do this at least 5500 times without degradation of optical contrast, switching speed or uniformity. Such dynamic window materials can be a promising alternative to those using traditional electrochromic materials.

Among alternative energy sources, photovoltaic devices play an important role. Nowadays, silicon-based solar cells, being the cheaper and most used, meet new demands to further increase their efficiency. Different approaches are proposed to enhance light absorption in a wide spectral range. For instance, nanostructured Si-based materials demonstrate high absorption ability due to quantum confinement effects (Gourbilleau et al. 2009). Silicon nanocrystals with the size of few nanometres (Figure 1.4, a) offer significant enhancement of light absorption and stimulate down- and upconversion processes. The application of Si nanocrystals in memory devices allows increases in the operation time and device stability that are required for long-term information storage.

Inorganic semiconductor nanostructures being placed on flexible substrates, such as natural and synthetic fibrous materials, or on rigid materials, such as ITO, glass and silicon

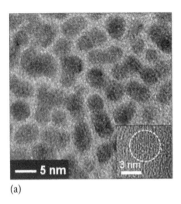
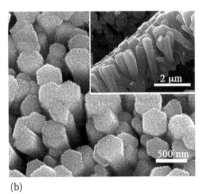

(a)   (b)

FIGURE 1.4   (a) Plan-view TEM image of solid state composite inorganic film with Si nanoparticles embedded in SiO_2 matrix. (Reproduced with permission from Gourbilleau et al., 2009. Copyright (2009) American Institute of Physics); (b) SEM image of ZnO/ZnS core-shell nanorods grown on cotton fibre and treated with Na_2S for 4 h. (Reproduced with permission from Athauda et al. 2014. Copyright 2014, The Royal Society of Chemistry.)

wafers, offer applications as UV photodetectors, photocatalysis, nanoscale lasers, protective and biocidal garments, etc. The coupling of inorganic nanostructures, for instance, ZnO-based ones, with flexible substrates (Figure 1.4b) is attractive for the development of flexible electronics, portable photovoltaic and solar cells. Their production is simple enough and environmentally friendly. Such materials can also find applications as stretchable strain sensors or as anode materials in flexible dye-sensitised solar cells.

It is worth pointing out that the composites mentioned above consist of inorganic constituents. However, there are the composites which are built as a mixture of inorganic and organic materials. Such composites are called *hybrid* materials or systems. The mixing of constituents in these materials occurs at the nanometre or molecular level.

Commonly, in hybrid materials the matrix is organic and the reinforcement is inorganic. In this case, such synergism produces materials whose properties are unavailable from the individual constituents. In most cases, the inorganic part provides mechanical strength and an overall structure to the natural objects, while the organic part delivers bonding between the inorganic building blocks and/or the soft tissue.

Organic–inorganic hybrid nanoflowers, a newly developed class of flower-like hybrid nanoparticles, have received much attention due to their simple synthesis, high efficiency and enzyme-stabilising ability. The fabrication of such nanomaterials is simplified by adding protein to metal ion solution; this synthetic method does not require any toxic elements or extreme harsh conditions. Therefore, the organic substance involved in the synthesis is subjected to less manipulation compared with other conventional methods to maintain the activity of the immobilised enzyme.

Figure 1.5 shows the reactions that occurred on the nanoflower surface. Compared with the corresponding free enzyme solutions, the enzyme efficiency of hybrid nanoflowers varies from 85 to over 1000%. These results suggest that the proteins in hybrid nanoflowers have higher activity and stability than the corresponding free enzyme solutions despite immobilisation in the flower petals. Moreover, hybrid nanoflowers overcome the

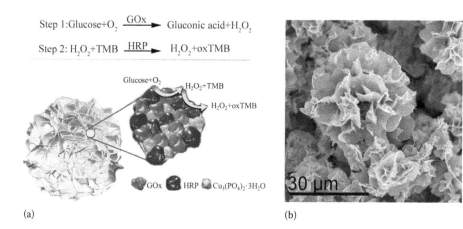

(a)   (b)

FIGURE 1.5 (a) The cascade enzymatic reaction of multi-enzyme co-embedded hybrid nano-flower for glucose detection. (Reproduced with permission from Sun et al. (2014). Copyright 2014, Royal Society of Chemistry); (b) GOx&HRP-$Cu_3(PO_4)_2 \cdot 3H_2O$ nanoflowers for glucose detection. (Reproduced from Lee, S. W. et al., *J Nanobiotechnol*, 13, 54, 2015, under the terms of the Creative Commons Attribution 4.0 International License (http://creativecommons.org/licenses/by/4.0/).)

previously encountered problem of mass-transfer limitation and open up an avenue for application to various research and detection fields and will lead to creative solutions and rapid development of biomaterials and biotechnology industries.

The development of composite and hybrid materials gives the designers flexibility in choosing an optimal combination of the original materials to build novel composites and hybrids within required parameters. Most complex shapes of the final product can be achieved.

The advantages of hybrid materials in comparison with solid state composites are the following: (1) inorganic clusters or nanoparticles with specific optical, electronic or magnetic properties can be incorporated in organic polymer matrices; (2) contrary to pure solid state inorganic materials that often require a high temperature treatment for their processing, hybrid materials show a more polymer-like handling, either because of their large organic content or because of the formation of cross-linked inorganic networks from small molecular precursors just like in polymerisation reactions; (3) light scattering in homogeneous hybrid material can be avoided, and therefore optical transparency of the resulting hybrid materials and nanocomposites can be achieved.

1.2 PHYSICAL METHODS FOR MANUFACTURING SOLID STATE COMPOSITES AND HYBRID SYSTEMS

1.2.1 Physical Vapour Deposition

This section gathers together different physical methods used for fabrication of solid state composites and hybrid systems. One of these methods is *physical vapour deposition* (PVD) that merges a variety of vacuum deposition approaches. It is used mainly for the production of thin films and coatings for mechanical, optical, chemical or electronic functions. The PVD process includes several steps: (1) the material to be deposited is converted into vapour by physical means; (2) the vapour is transported across a region of low pressure from its source to the substrate; (3) the vapour undergoes condensation on the substrate to form the thin film. There are many parameters that control sputter deposition, making it a complex process, but their monitoring allows experts a large degree of control over the growth and microstructure of the film.

The most common PVD processes are *sputtering* and *evaporation*.

In *PVD by evaporation*, the conversion into vapour phase is achieved by applying heat to the source material, causing it to undergo evaporation. The vaporised atoms or molecules are transported to the substrate in a high-vacuum environment allowed minimal collision interference from other gas atoms or molecules. The rate of mass removal from the source material as a result of such evaporation increases with vapour pressure, which in turn increases with the applied heat. Vapour pressure greater than 1.5 Pa is needed in order to achieve deposition rates which are high enough for manufacturing purposes.

The advantages offered by evaporation for PVD are: (1) high film deposition rates; (2) low substrate surface damage from impinging atoms as the film is being formed; (3) excellent purity of the film because of the high vacuum condition used by evaporation; (4) less tendency for unintentional substrate heating. The disadvantages of this method are: difficult control of film composition in the case of composite films; absence of capability to do *in situ* cleaning of substrate surfaces; step coverage is more difficult to improve by evaporation; X-ray damage caused by electron beam evaporation can occur.

In the semiconductor industry, PVD by evaporation is used primarily in the deposition of aluminium and other metallic films on the wafer.

The *PVD by sputtering* can be described as a sequence of the next steps: (1) ions are generated and directed at a target material; (2) the ions sputter atoms from the target; (3) the sputtered atoms get transported to the substrate through a region of reduced pressure; (4) the sputtered atoms condense on the substrate, forming a thin film.

Sputtering yield, or the number of atoms ejected per incident ion, is an important factor in sputter deposition processes, since it affects the sputter deposition rate. Sputtering yield primarily depends on three major factors – target material, mass of the bombarding particles and energy of bombarding particles. In the energy range where sputtering occurs (10–5000 eV), the sputtering yield increases with particle mass and energy.

Sputtering offers the following advantages over other PVD methods: (1) sputtering can be achieved from large-size targets, simplifying the deposition of thin films with uniform thickness over large wafers; (2) film thickness is easily controlled by fixing the operating parameters and simply adjusting the deposition time; (3) control of the alloy composition, as well as other film properties such as step coverage and grain structure, is more easily accomplished than by deposition through evaporation; (4) sputter-cleaning of the substrate in vacuum prior to film deposition can be done; (5) device damage from X-rays generated by electron beam evaporation is avoided. However, this method has the following disadvantages too: (1) high capital expenses are required; (2) the rates of deposition of some materials (such as dielectrics) are relatively low; (3) some materials, such as organic solids, are easily degraded by ionic bombardment; (4) sputtering has a greater tendency to introduce impurities in the substrate than deposition by evaporation because the former operates under a lower vacuum range than the latter.

The development of the sputtering methods during the last few decades shows that some disadvantages of the PVD by sputtering can be omitted. The simplest sputtering approach is the direct current (DC)-powered diode. However, it has turned out to be unsuitable for many industrial applications since the deposition rates are low and the thermal load to substrates is high. It is not possible to deposit films on temperature sensitive materials like plastic discs or foils. Since the introduction of the planar magnetron by J.S. Chapin in 1974 (the patent was issued in 1979), magnetron sputtering has become the most important technology for the deposition of thin films.

Nowadays, all industrial branches need high-quality coatings for the realisation of new or the improvement of existing products. The magnetron cathode combines the advantages of economic deposition even on large areas and the ability to coat very temperature sensitive plastic substrates. The main problems, like poor target material utilisation of the planar magnetron or process instabilities during deposition of highly insulating films, have been solved by many innovations during the last decades. Moreover, novel approaches such as "High Power Impulse Magnetron Sputtering (HiPIMS)", "Sputter Yield Amplification (SYA)" or sputtering from hot targets were recently developed.

1.2.2 Magnetron Sputtering

As it was mentioned above, *magnetron sputtering* belongs to the physical vapour deposition process. It is highly versatile and fully compatible with established complementary metal-oxide-semiconductor (CMOS) technologies. Sputtering sources are equipped with magnetrons that utilise strong electric and magnetic fields to confine charged plasma particles close to the surface of the sputter target. In a magnetic field, electrons follow helical paths around magnetic field lines, undergoing more ionising collisions with gaseous neutrals near the target surface. The sputter gas is typically argon. The sputtered atoms are neutrally charged and so are unaffected by the magnetic trap. Charge build-up on insulating targets can be avoided with the use of *RF sputtering* where the sign of the anode-cathode bias is varied at a high rate (commonly 13.56 MHz). RF sputtering works well to produce highly insulating oxide films.

The main advantages of magnetron sputtering are as follows: (1) low plasma impedance and thus high discharge currents from 1A to 100A (depending on cathode length) at typical voltages around 500V; (2) deposition rates in the range from 1 nm/s to 10 nm/s; (3) low thermal load to the substrate coating uniformity in the range of a few per cent even for several meters-long cathodes; (4) easy to scale up; (5) dense and well adherent coatings; (6) large variety of film materials available (nearly all metals, compounds, dielectrics); (7) broadly tuneable film properties.

There are a lot of modifications of this method. However, all of them can be gathered in two main approaches: (1) sputtering, when the film is grown from the sputtering of a single target (either simple or composite material); (2) co-sputtering, when the composite film is grown via simultaneous sputtering of several targets (Figure 1.6).

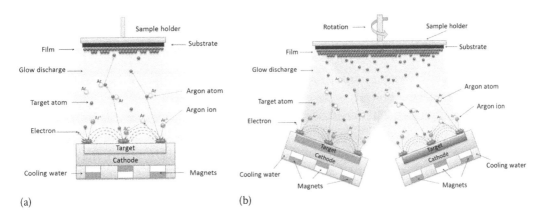

(a) (b)

FIGURE 1.6   A schematic illustration of magnetron sputtering from single cathode (a) and confocal co-sputtering from two spaced-apart cathodes (b).

In the case of sputtering, the physical size of the target exceeds the size of the substrate, while the deposition can be done on the non-rotated substrate. Co-sputtering is usually performed on the rotated substrate to achieve homogeneous films. The co-sputtering is more flexible. The deposition can be performed on the large substrate. Besides the growing of thick layers, it also allows an easy fabrication of superlattice structures.

The sputtering can be performed in the plasma stimulated by Ar ions only (called "standard deposition/sputtering") or in the mixture of Ar and N_2 (or O_2, or H_2) (reactive sputtering). Nowadays, the fabrications of composite thin films via simultaneous sputtering of metallic, semiconductor and/or dielectric targets can be also achieved.

1.2.3 Ion Beam Implantation

Ion implantation is one of the fundamental processes of CMOS technology. This approach is used in semiconductor device fabrication and in metal finishing, as well as in materials science research. The ions can alter the elemental composition of the target (if the ions differ in composition from the target) when they stop and remain in the target. By ion implantation, when the ions impinge on the target at high energy, chemical and physical properties of materials can be changed. Advantages of ion implantation are: (1) precise control of dose and depth profile of impurities; (2) low-temperature process that can use photoresist as a mask; (3) wide selection of masking materials, e.g. photoresist, oxide, poly-Si, metal; (4) Less sensitive to surface cleaning procedures and (5) excellent lateral dose uniformity (< 1% variation across 12" wafer). Besides, ion implantation is a versatile and powerful technique for forming many types of nanocrystalline precipitates embedded in the near-surface region of a wide variety of crystalline and amorphous host materials. However, this method has some disadvantages. One of them is the nonuniform distribution of the implanted ions and the projectile-induced structural damage. However, these disadvantages can be compensated by using multi-energy ion implantation and subsequent annealing processes. Compared to magnetron sputtering, the ion implantation is less flexible. An implantation is only possible near the surface. The fabrication of superlattice structures is thus impossible.

1.3 CHEMICAL METHODS FOR MANUFACTURING SOLID STATE COMPOSITES AND HYBRID SYSTEMS

1.3.1 Colloidal and Sol-Gel Processes

As it was already written, traditionally composite materials are classified according to their matrix materials – as ceramic, metal and polymer Matrix composites (Camargo et al. 2009). The most common chemical techniques for the processing of ceramic and metal matrix nanocomposites include colloidal and sol-gel processes. These methods are simple, versatile and effective low-temperature with high chemical homogeneity and high purity products. For example, colloidal CdSe and CdSe/CdS core-shell nanocomposites capped with the molecular metal chalcogenide complex In2Se42- for photovoltaics application were presented in Lee et al. (2011). The photodetector with normalised detectivity $D^* > 1 \times 10^{13}$ cm $Hz^{1/2}$ W^{-1}, which is very good for II-VI nanocrystals, was designed. Invention (Nag and Talapin 2014) describes a method of nanocomposite fabrication from the isolable colloidal particle comprising a nanoparticle and an inorganic capping agent bound to the surface of the nanoparticle. This method can yield photovoltaic cells, piezoelectric crystals, thermoelectric layers, optoelectronic layers, light-emitting diodes, ferroelectric layers, thin film transistors, floating gate memory devices, phase change layers and sensor devices. Invention (Chan et al. 2017) provides new, efficient synthetic strategies of a mesoporous nanomaterial by using colloidal solution combustion synthesis wherein the colloidal nanoparticles have uniform arrangement in the nanocomposite produced and the nanocomposite can be composed of oxide/oxide, oxide/metal or metal/oxide possessing a unique nanostructure.

A good example of a new class of colloidal core-shell materials prepared with multiple "spikes" formed from the self-assembly of gold nanorods (AuNRs) onto a spherical polystyrene (PS) core is presented by Guo et al. (2014). The assembled structures could serve as a versatile substrate for surface-enhanced Raman spectroscopy-based sensing applications (see Figure 1.7).

The sol-gel process developed in the 1930s is one of the major driving forces behind what has become the broad field of inorganic-organic hybrid materials. The review of literature reveals a high potential of sol-gel chemistry in the area of producing a wide range of composite and hybrid structures (Pandey and Mishra 2011; Kumar et al. 2015). Its essence is in the formation of an oxide network through polycondensation reactions of a molecular precursor in a liquid. From the basic concepts, there are many possibilities for incorporating organics into sol-gel matrices, for example, to bind organic groupings covalently to inorganic network formers such as the silicon- tetrahedron by using organoalkoxysilanes as sol-gel precursors. Recent studies have shown significant potential of the sol-gel method in biomedical areas (Owens et al. 2016).

At the same time, the weak bonding, low wear-resistance and difficult control of porosity can be attributed to the limitations of the named chemical methods (Camargo 2009). It should also be noted that solution-based chemical techniques have some limitations, primarily associated with limited stability of reactants and solvents at the high temperatures required for some hard-to-crystallise nanomaterials. This fundamental problem can

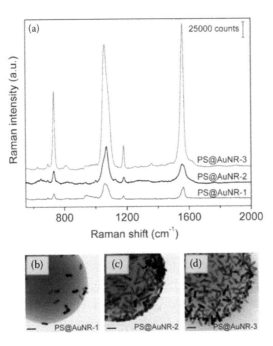

FIGURE 1.7 Colloidal core-shell materials with "spiky" surfaces assembled from gold nanorods: (a) Surface-enhanced Raman spectra of 1,4-benzenedithiol (1,4-BDT) adsorbed onto three different samples of PS@AuNRs with an increasing loading of AuNRs on the PS core. (b–d) Representative TEM images of PS@AuNRs assemblies depicting the three different loadings of AuNRs, respectively. Scale bars in each TEM image are 50 nm in length. (Reproduced with permission from Guo et al. (2014). Copyright 2014, The Royal Society of Chemistry.)

be addressed by moving away from solution chemistry and employing solid state reactions or synthesising nanoparticles in the gas phase.

1.3.2 Photochemical and Electrochemical Methods

Along with chemical methods, the use of photochemical approach to the formation of nanocomposites is very effective (Stroyuk et al. 2008). The photochemical method of synthesis has proven more effective than the production of nanocomposite by traditional synthetic approaches, and in a number of cases only they make it possible to achieve the assigned aim. The photochemical production of metal compounds (such as Sulphides, Selenides and Tellurides) is based on the photodecomposition of complexes in which the chalcogen and metal have been included simultaneously or on the photoreduction of sulphur- or selenium-containing compounds in the presence of salts of the metals. The next step to the photochemical formation of nanocomposites is the photodeposition of the nanoparticles of sulphides on the surface of other semiconductors. Thus, such structures as TiO_2/MoS_2, TiO_2/WS_2, $TiO_2/CdSe$ and $TiO_2/PbSe$ are obtained. The photocatalytic reduction of elemental sulphur in alcohol media with the participation of ZnO nanoparticles and nanostructured films of TiO_2 in the presence of metal salts was used for the formation of a series of binary nanocomposites, ZnO/MS (Shvalagin et al. 2007) and TiO_2/MS (MS = CdS, PbS, CuS), and the reaction with the participation of a TiO_2/Au^0 nanocomposite in the presence of Cd^{II} was used for the formation of a ternary TiO$_2$/Au/CdS heterostructure in which cadmium sulphide was deposited on the surface of gold nanoparticles in the form of a layer 1–2 nm thick (Tada et al. 2006).

We will show a variety of chemical methods on an example of only one composite, MoS$_2$/TiO$_2$. The synthetic strategies of this composite can be roughly divided into two categories: *ex situ* synthetic strategies and *in situ* synthetic strategies (Chen et al. 2018). In the *ex situ* synthetic strategy, MoS$_2$ and TiO$_2$ are prepared separately in advance; then, the MoS$_2$/TiO$_2$ composites are synthesised by drop-casting, impregnation, ball milling, sol-gel, or hydrothermal/solvothermal assembly methods. For the *in situ* synthetic strategy, the synthesis process involves ionic reactions; this results in more uniform dispersion and complex interactions compared with the *ex situ* synthetic strategy (Chen et al. 2018).

Another method about which we will write is electrochemical synthesis. This is the synthesis of chemical compounds in an electrochemical cell. The main advantage of electrosynthesis over an ordinary redox reaction is avoidance of the side reactions and the ability to precisely tune the required power input to achieve the chemical reaction. Compared to gas phase depositions (e.g. atomic layer deposition and organometallic vapour phase epitaxy), which require specific vacuum and/or heating conditions as well as the use of specialised precursors, solution-based electrodeposition performed under ambient conditions is inexpensive and easily scalable. This makes electrodeposition more attractive for producing semiconductor and catalyst electrodes for use in energy-related applications.

The relevance of this method can be confirmed by the fact that it is applied for producing graphene by cleavage of graphene by embrittlement from the graphite anode forming graphene/graphene oxide composite (Santhanam et al. 2016). Another invention demonstrates the preparation of polymer/oxygen-free graphene composites through an electrolytic exfoliation process (Tuantranont et al. 2012). The unique property of produced graphene is oxygen-free, while the polymer-graphene composites are demonstrated to be highly transparent and conductive. The invention also includes the use of polymer-graphene composite produced by the said method as an ink for inkjet printing to deposit electrically conductive layers on a given substrate for a variety of applications including transparent conductors, transparent antenna and electrochemical electrodes.

1.3.3 Sonochemical and Hybrid Methods

As it is known, the phenomenon of acoustic cavitation lies at the basis of sonochemical synthesis of new materials. The creation of complex structures on the solid surface or in the liquid media involving cavitation phenomenon occurs due to the generation of microjets and shock waves during the collapse of cavitation bubbles and stipulated by the chemical effect of the ultrasound based on the generation of free radicals during cavitation. According to researchers working in the field of fabricating new nanostructured materials, the obvious advantages of the ultrasound-assisted methods over the traditional procedures are presented by more uniform size of the nanoparticles distribution, a higher surface area, faster reaction time and improved phase purity. Applying the cavitation effects allows accelerating a reaction or permitting the use of less aggressive conditions, reducing the number of steps which are required using normal methodology, and opens the possibilities for alternative reaction pathways (Bang and Suslick 2010).

The incredible possibilities of sonochemical methods make it possible to create nanostructured solids of various forms and morphologies with unique properties for different applications. The ultrasound demonstrates multiform and, simultaneously, complex effects on solid state, which consists in solid surface modification and exfoliation of layered materials; particles agglomeration/fragmentation and new materials fabrication by deposition of nanoparticles onto different surfaces, including mesoporous materials; core/shell structure and polymer-based hybrid nanocompounds fabrication. Sonochemistry advances can be summarised in the block diagram presented in Figure 1.8.

The review of literature and patents reveals a high potential of ultrasound-assisted approaches in composite and hybrid structure manufacturing. Invention (Cobianu et al. 2016) describes a method for forming a hydrogen sulphide sensor developed using a simple sonochemical route that includes obtaining a metal oxide heterostructure (CuO/ZnO or CuO/SnO$_2$) from a batch sonochemical synthesis. During sonication, an inert ambient can be kept above the solution, which can reduce or prevent carbon dioxide from contacting or reacting with the solution so as to reduce or prevent metal carbonates' formation during sonochemical synthesis.

Ultrasound was applied during a gas sensor fabrication (Cobianu et al. 2017). A sensor for room temperature detection of ethanol (C$_2$H$_5$OH) vapours utilises a high-sensitivity field-effect transistor and nanostructured inorganic-organic nanofibers, prepared by means of a combined sonochemistry-electrospinning method.

The deposition of metal oxides known to possess antimicrobial activity, namely ZnO, MgO and CuO, can significantly extend the applications of textile fabrics, medical devices and other items and prolong the period of their use. Invention of (Gedanken et al. 2016) describes

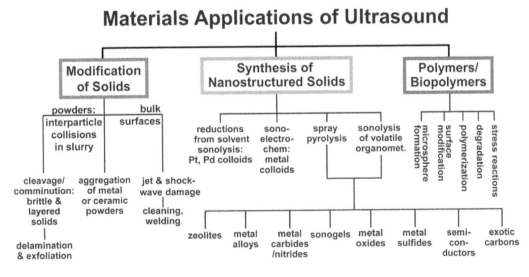

FIGURE 1.8 Block diagram summarised multiplicity of the sonochemistry applications to materials synthesis and modification. (Adapted from Xu et al. (2013), with permission. Copyright 2012, The Royal Society of Chemistry.)

sonochemical coating of textiles with metal oxide nanoparticles for antimicrobial fabrics. This new concept was developed for the formation of human- and environmentally friendly materials able to provide antibacterial activity. Functionalised hydroxyapatite/gold composites as "green" materials with antibacterial activity and the sonochemical method, which has been developed for composite preparing, was patented in Vukomanovic et al. (2015).

It should be noted that the coupling of ultrasound with other chemical methods, such as electrochemistry or photochemistry, has resulted in new powerful methodologies, which enhance the mechanical properties of the deposits and electrocatalytic performance, in the case of sonoelectrodeposition. In sonophotodeposition, ultrasound influence is reflected in the enhanced photochemical reaction rates but can also end in a modified reaction pathway (Magdziarz and Colmenares 2017). The combination of ultrasound-based synthesis with microwave heating has been also successfully exploited in applied chemistry. Besides saving energy, these green techniques promote faster and more selective matter transformations (Martina et al. 2016).

The sonochemistry success in the field of creation of composite and hybrid structures will be described in Chapter 4.

1.4 ANALYTIC APPROACHES FOR MATERIALS CHARACTERISATION

1.4.1 Fourier Transform Infrared Spectroscopy

This method is an effective analytical technique for quickly identifying the "chemical family" of a substance. FTIR measures the absorbance of infrared light by a sample and generates a spectrum based on the functional groups in the material.

Typically, organic and polymeric compounds produce a "fingerprint" IR spectrum, which can be compared to an extensive reference database, and the unknown component's chemical family or actual identity may be determined. In the case of inorganic materials, FTIR spectra are used to a lesser degree. At the same time, they allow the information about microstructure transformation on composite materials sensitive to IR excitation to also be obtained in an express and non-destructive way. One of the bright examples of the utility of FTIR method is the estimation of excess Si content in Si-rich-SiO_2 and Si-rich-Si_3N_4 composite materials, investigation of their microstructural evolution with annealing treatment followed by the formation of Si nanocrystals (see for instance, Khomenkova et al. 2013; Debieu et al. 2013) (Figure 1.9). The main principle of this method is based on the comparison of the parameters (such as peak position, full-width and intensity) of specific vibration bands of composite material with those of a pure dielectric host.

In the case of Si-rich-SiO_2 materials, the TO_3 Si-O phonon mode is the most sensitive mode for both the bonding configuration and the SiO_x composition, suggesting that it is made up from sub-bands with variable relative weight. This band consists of two sub-bands centred at 1054 and 1090 cm^{-1} corresponding to Si-O-Si bond angles of 131° and 143°, respectively (Lisovskii et al. 1995), the 1050 cm^{-1} band having an increased weight in SiO_2 films of higher density. In the case of Si-rich SiO_2 composites (called also as SiO_x), the TO_3 peak contains four major sub-bands with peak positions at 1000, 1038, 1066 and 1092 cm^{-1} corresponding to the several ($Si-O_{4-k}-Si_k$, $0 \leq k \leq 4$) bonding tetrahedra to which the Si next neighbours of the vibrating O-atom may belong (Figure 1.9a,b). The band at

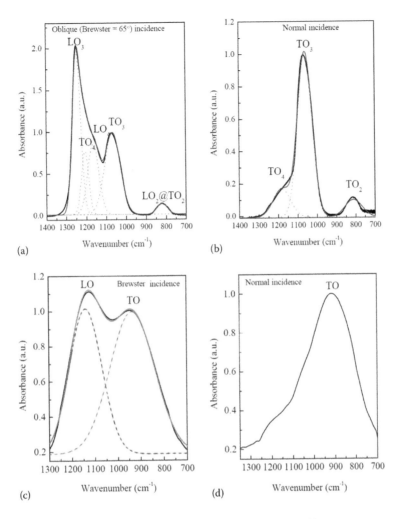

FIGURE 1.9   FTIR spectra of SiO_2 (a, b) and Si_3N_4 (c, d) films produced by magnetron co-sputtering and recorded at oblique (Brewster angle, 65°) incidence (a, c) and normal incidence (b, d). The decomposition the spectra on corresponding TO and LO phonon modes is also shown.

1092 cm^{-1} is assigned to the Si-(O_4) tetrahedron, dominating the spectrum in case of thermally grown or annealed SiO_2 films. The formation of Si-NCs during thermal annealing of nonstoichiometric SiO_x leads to a higher fraction of Si-O-Si configurations in which the two Si next neighbours of the vibrating O-atom are part of either Si-(O_4) or Si-(O_3Si) bonding tetrahedral. As a consequence, the relative weight of the 1092 cm^{-1} sub-band increases upon SiO_x annealing and results in the shift of TO_3 peak position towards that of pure SiO_2 (Schmidt and Schmidt 2003) (Figure 1.10). The higher-energy shift of LO_3 and TO_3 peak positions as well as quenching of LO_4-TO_4 modes testifies to the phase separation process via "decreasing" of excess silicon distributed randomly in the composite.

The higher difference between TO_3 peak position of SiO_2 and SiO_x materials, the more excess Si content is incorporated into SiO_2 host. Based on the approach proposed in (Cueff et al. 2011; Liang et al. 2012 and Khomenkova et al. 2013), the excess Si content can be extracted for SiO_x composites. As for Si-rich-Si_3N_4 composited, the utility of FTIR method for Si excess evaluation was also demonstrated (Figure 1.9c,d) (see, for instance, Debiue et al. 2013).

One can point out that applying the same procedure of the comparison of TO_3 mode peak position observed for annealed samples, it is possible to follow the progress of excess Si conversion into Si nanoclusters at mediate temperatures and into nanocrystals at high annealing temperatures (Figure 1.10). The appearance of Si crystallites is usually reflected by the high intensity of LO_3 Si-O phonon that is proved to be a feature of Si/SiO_2 interface of a good quality. More details about the application of FTIR method to characterise different composites can be found in Chapters 4 and 6 of this book.

1.4.2 Spectroscopic Ellipsometry

Ellipsometry and, in particular, spectroscopic ellipsometry are very capable optical methods used for composites' characterisation due to providing numerous advantages. The high sensitivity combined with the high speed and the non-destructive nature results in two main capabilities: (1) the sample properties can be quickly and non-destructively qualified as well as mapped on large surfaces; (2) based on the non-destructive nature and the *in situ* application, there is a possibility of following the transformation of the sample structure during deposition and/or annealing. Even though such optical measurements can be performed, a suitable analysis can be completed only if proper optical models have been developed and accurate reference dielectric functions exist.

The optical characterisation of composite materials is very challenging since the samples contain mixed phases, and the dielectric functions of the constituents are changing with the annealing conditions and stack-depth. To determine the volume fraction of the phases as well as to follow the phase separation and the changes in the crystalline structure of the phases accurately, the Bruggeman effective medium approximation is usually applied. Among different recent demonstrations of the utility of this analytical tool, one

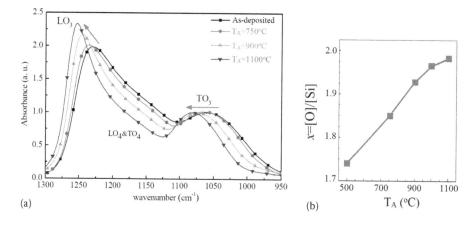

FIGURE 1.10 (a) Evolution of FTIR spectra of Si-rich SiO_2 composite film with annealing recorded at oblique (Brewster angle, 65°) incidence; (b) variation of the factor stoichiometry $x = [O]/[Si]$ with annealing temperature.

can refer to the report of Agocs et al. (2017) on the characterisation of Ge-rich-ZrO_2 composite materials and superlattices on their basis. Using effective medium approximation, as well as applying the dielectric functions for Ge and ZrO_2 components described by the Cauchy and the Urbach-Cody-Lorentz parameterisation, respectively, the fitting of the experimental data for as-deposited materials was performed with high accuracy (below 1%) (Figure 1.11).

An annealing of these composites results in the significant transformation of experimental spectra (Figure 1.12a), however, the depolarisation effect is negligible. It was assumed that such transformation of the spectra was caused not only by the phase separation process but rather significant variation of chemical composition of the material, especially in the near-surface region.

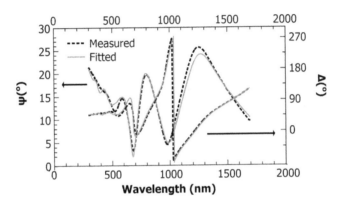

FIGURE 1.11 Measured and fitted (with single-layer model) spectra for the as-deposited Ge-rich-ZrO_2 composite film. The angle of incidence was 70°. (Reproduced with permission from *Appl Surf Sci*, 421, Agocs et al., Optical and structural characterisation of Ge clusters embedded in ZrO_2, 283–288, Copyright 2017, Elsevier.)

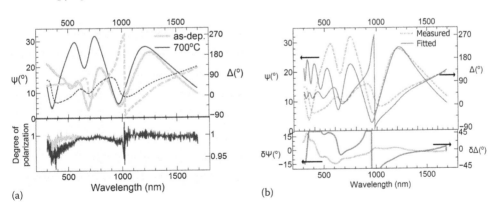

FIGURE 1.12 (a) Measured and fitted spectra using the single-layer model for Ge-rich-ZrO_2 composite film annealed at 700°C. The angle of incidence was 70°. (b) The comparison of experimental data for as-deposited and annealed films that demonstrate the negligible role of polarisation effect. (Reproduced with permission from *Appl Surf Sci*, 421, Agocs et al., Optical and structural characterization of Ge clusters embedded in ZrO_2, 283–288, Copyright 2017, Elsevier.)

This was evident by the results obtained for the composites annealed at 700°C applying the single-layer model with vertical homogeneity (Figure 1.12b).

The fit quality deteriorated for these temperatures with mean square error values jumping from 50 to over 200. It was supposed that the variation of chemical composition over film bulk occurred, and the multilayer model describing the gradual variation of dielectric functions was applied, allowing a minimisation of the mean square error. It was found that the surface region (several layer-pairs from the top) revealed substantial Ge out-diffusion. The out-diffusion of germanium from the top-surface region was confirmed by the Auger electron spectroscopy.

1.4.3 Raman Scattering Spectroscopy

One more optical method is Raman scattering spectroscopy (or Raman analysis). It enables the determination of chemical structure and the identification of compounds using vibrational spectroscopy. Raman scattering method has better spatial resolution than the FTIR method described above, allowing the analysis of smaller dimensions, down to the 1 μm range. This method is an ideal technique for the qualitative analysis of organic and/or inorganic mixed materials and can also be employed for semi-quantitative and quantitative analysis. It is often used to identify organic molecules, polymers, biomolecules and inorganic compounds both in bulk and in individual particles. For imaging and depth profiling it is used to map the distribution of components in mixtures, such as drugs in excipients, tablets and drug-eluting stent coatings. Raman scattering allows determining of inorganic oxides and their valence state as well as the stress and crystalline structure in semiconductors and other materials. It can be used, for instance, to determine the amorphous/crystalline phase contribution in mixed materials as well as to estimate the mean size of the nanocrystalline entities. Some applications of this method, for instance, for the estimation of the sizes of semiconductor nanocrystals, are widely addressed. Let's show the utility of this method for evaluation of the Si nanocrystals appearing due to phase separation in SiO_x or SiN_x materials described above. When Si nanocrystals approach nanometre scale, the corresponding Raman scattering spectra show their broadening and shift of peak position to lower wavenumber side in comparison with that of bulk materials.

For the analysis of Raman scattering spectra, several models were proposed. Among them, phonon confinement model (PCM) and bond polarisability model (BPM) are the most addressed. Figure 1.13 shows the evolution of the shape of Raman scattering spectra versus the size of Si nanocrystals calculated with the PCM model as well as the estimated mean size versus the spectral shift of peak position. As one can see, this method is sensitive to the little variation of crystallite size. Besides, it is also sensitive to the stresses in the materials. Based on the comparison of experimental data with theoretical data, one can extract information about the properties of nanocrystals. This method is an express and non-destructive one. Its application for the analysis of nanostructured composite materials will be demonstrated in the following chapters.

1.4.4 Luminescence

Photoluminescence spectroscopy is a non-destructive approach for the investigation on electronic structure of materials. It allows one to obtain important information about

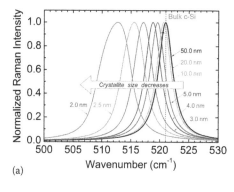
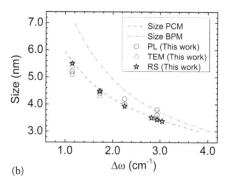

(a) (b)

FIGURE 1.13   (a) Calculated Raman spectra of silicon nanocrystals with various sizes by employing the phonon confinement model (PCM); (b) Change in the peak size of small Si-NCs as a function of the absolute Raman peak shift to the lower wavenumbers with respect to the peak position of bulk c-Si (521 cm^{-1}) according to our approach based on the PCM. Raman scattering (RS), photoluminescence (PL) and transmission electron microscopy (TEM) data are represented as stars, circles and triangles, respectively. Each Raman measurement (stars) from a sample is accompanied by its corresponding PL and TEM data in the vertical direction. For comparison with PCM model, the BPM one is also demonstrated in the plot. (Reprinted with permission from Dogan et al. (2013). Copyright 2013, American Institute of Physics.)

excitation, relaxation and emission mechanisms. The excitation mechanism lies in the absorption of the photons by the material leading to excitation of an electron from ground state to the allowed excited state. In the case of semiconductors, this excitation process occurs between the valence and the conduction bands. This process is characterised by the rise time (τ_{rise}) that depends on the photon flux and their energy.

The relaxation of excited carriers can occur in two ways, i.e. via a non-radiative or radiative recombination. The former allows the electron to go back to the ground state by transferring its energy to the network of the material via several processes (phonon assisted, Auger process, etc.). The latter process is accompanied by the emission of a photon. The schematic presentation of luminescence phenomena is shown in Figure 1.14.

Usually, the emission process is characterised with a decay time (τ_{decay}), which takes into account the radiative and non-radiative recombination. The non-radiative process acts as a competing process resulting in somehow quenching the emission. In this case, the decay time can be described as $1/\tau_{decay} = 1/\tau_{radiative} + 1/\tau_{non\text{-}radiative}$ and characterises the time of electron relaxation form conduction to valence band.

It should be noted that among different optical methods, luminescent methods (such as photoluminescence, photoluminescence excitation and cathodoluminescence) play significant roles in the investigation of different composite and hybrid materials. The specificity of luminescent methods and their application to the analysis of material properties will be demonstrated in most chapters of this book.

1.4.5 Auger Electron Spectroscopy

The *Auger electron spectroscopy* technique (named after Pierre Auger, who described this process in 1925) uses a primary electron beam, typically in the 3–25 keV range. Atoms

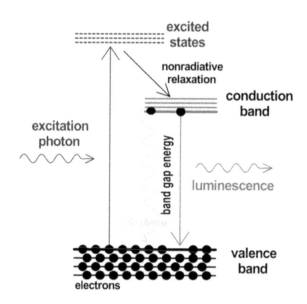

FIGURE 1.14 Schematic presentation of luminescence process.

that are excited by the electron beam can relax through the emission of Auger electrons. The kinetic energies of the emitted Auger electrons are measured and are characteristic of elements present at the surface of a sample. The resulting spectrum is usually plotted as the derivative of the signal intensity versus kinetic energy, with each element showing a unique "fingerprint" for elemental identification. Auger's high spatial resolution capabilities and surface sensitivity make it the technique of choice for the following types of applications: analysing sub-µm particles to determine contamination sources; identifying defects in electronic devices to investigate failure causes; determining the oxide layer thickness of electro-polished devices; mapping elemental distribution on discoloured or corroded regions; cross-sectional analysis of buried defects in film stacks; identifying grain boundary contamination in metal fractures, fatigues and failures; integrity and uniformity of thin film coatings. This approach has high spatial resolution: <10 nm minimum spot size; surface sensitive (top 5–10 nm) with accuracy of 0.1 at % to 1 at %; identification of all elements except H and He; 2D and 3D elemental distribution of small areas; rapid analysis for elemental composition; and can analyse up to 300 mm wafers. To demonstrate the utility of this approach, one can refer some recent experiments on SiGe composite nanocrystals (Ponomaryov et al. 2016) where an application of scanning Auger microscopy with ion etching technique and effective compensation of thermal drift of the surface analysed area was proposed. This method allowed direct local study of composition distribution in the bulk of single Ge and/or Ge_xSi_{1-x} nanoislands grown by molecular beam epitaxy on Si substrate (Figure 1.15).

Figure 1.15a represents the plan view of the sample obtained by scanning electron microscopy where formed Ge nanocrystals are well-distinguished. The analysis of their composition (Figure 1.15b–d) shows the concentration profile, in which one can clearly see three zones: the Si-cap layer, Si-shell penetrating into nanoisland to a depth of 5–7 nm and Ge-core. The interdiffusion mixing in the capped and open (after ion etching) nano-structures is seen. Lateral distributions of the elemental composition as well as concentration-depth profiles show that there is a germanium core inside the nanocrystals that even penetrates into the Si substrate. The highest Ge content was found to be about 40 at % allowed to determine the Si/Ge depth distribution in SiGe nanocrystal if the size is about 30 nm. The resolution of this method and its accuracy allows precise determination of the chemical composition of the composite materials at nanoscaled level.

1.4.6 X-ray Diffraction and X-ray Reflectivity

X-ray diffraction (XRD) is a powerful non-destructive technique for characterising crystalline materials. It provides information on crystal structure, crystalline phase (polymorphs), preferred crystal orientation (texture) and other structural parameters, such as average grain size, crystallinity, strain and crystal defects. XRD peaks are produced by the constructive

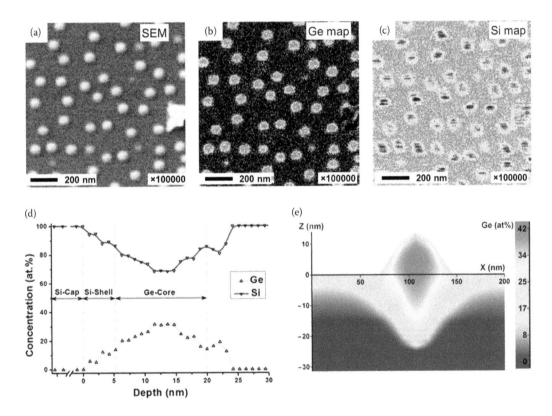

FIGURE 1.15 (a) SEM image and (b, c) the Auger electron maps for Ge (b) and Si(c) distribution over the surface of the sample with SiGe nanocrystals; (d) concentration-depth profile, which begins with Si-cap layer and stretches across Si-shell, Ge-core up to Si substrate; (e) an axial section of 3D distribution of Ge content in SiGe nanocrystal. (Reproduced from *Nanoscale Res Lett,* Direct determination of 3D distribution of elemental composition in single semiconductor nanoislands by scanning auger microscopy, 11, 2016, 103, Ponomaryov et al., Figure 1.15, Copyright Springer under the terms of the Creative Commons Attribution 4.0 International License (http://creative-commons.org/licenses/by/4.0/).)

interference of a monochromatic beam of X-rays diffracted at specific angles from each set of lattice planes in a sample. The peak intensities are determined by the distribution of atoms within the lattice. Consequently, the X-ray diffraction pattern is the fingerprint of the periodic atomic arrangements in a given material. A search of the ICDD (International Centre for Diffraction Data) database of X-ray diffraction patterns enables the phase identification of a large variety of crystalline samples. The main applications of XRD analysis are: identification/quantification of crystalline phase; measurement of average crystallite size, strain or micro-strain effects in bulk materials and thin film; quantification of preferred orientation (texture) in thin films, multilayer stacks and manufactured parts; determination of the ratio of crystalline to amorphous material in bulk materials and thin-films. Some examples of the utility of the XRD method for film characterisation are shown in Figure 1.16a. The effect of annealing on the structure of thin HfO_2 film grown on Si substrate can be easy distinguished. The threshold temperature for film crystallisation is above 800°C. Another application of XRD method will be demonstrated, for instance, in Chapters 2 and 6 of this book.

Another XRD-related technique is *X-ray reflectivity (XRR) that is widely used for the determination of thin-films and multilayer structures.* X-ray scattering at very small diffraction angles allows characterisation of the electron density profiles of thin films down to a few nanometres thick.

Using a simulation of the reflectivity pattern, a highly accurate measurement of thickness, interface roughness and layer density for crystalline and amorphous thin films and multilayers can be obtained. Besides, it allows fine control of the quality of multilayers, especially annealed ones, in terms of interdiffusion and losing periodicity. No prior

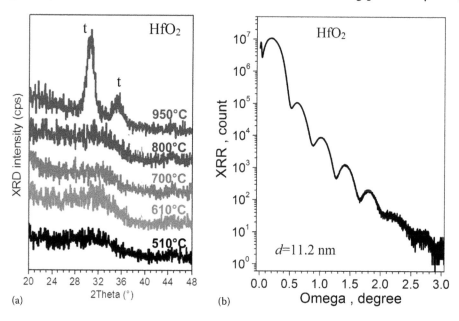

(a)

(b)

FIGURE 1.16   (a) Typical XRR spectrum for as-deposited thin HfO_2 film grown on Si substrate by magnetron sputtering. The thickness was estimated to be about 11.2 ± 0.1 nm. (b) Evolution of XRD patterns of the same HfO_2 film with annealing at different temperatures. The transition from amorphous to tetragonal crystalline phase occurs at temperature higher than 800°C.

knowledge or assumptions regarding the optical properties of the films are required, unlike optical ellipsometry.

1.4.7 Electron Microscopy

Transmission electron microscopy (TEM analysis) and scanning transmission electron microscopy (STEM) are similar techniques that image a sample using an electron beam. Image resolutions are around 1–2Å for TEM analysis and STEM. High-energy electrons (80 keV–200 keV) are transmitted through electron transparent samples (~100 nm thick). TEM and STEM have better spatial resolution than scanning electron microscopy (SEM) but often require more complex sample preparation. Though TEM and STEM are more time-intensive than many other frequently used analytical tools, a variety of signals are accessible from these techniques, making it possible to perform chemical analysis at the nanoscale. Besides high image resolutions, it is possible to characterise crystallographic phase and crystallographic orientation (using electron diffraction experiments), generate elemental maps (by using energy dispersive spectroscopy, for instance) and acquire images highlighting elemental contrast (Z-contrast or High-Angle Annular Dark-Field Scanning Transmission Electron Microscopy [HAADF-STEM] mode). These can all be accomplished from precisely located areas at the nm scale. STEM and TEM are excellent failure analysis tools for thin film and nanostructured materials (Figure 1.17).

TEM methods allow the identification of nm-sized defects on integrated circuits, including embedded particles and via residues; determination of crystallographic phases at the nanometre scale; nanoparticle characterisation (size, agglomeration, interface nature, effects of annealing, etc.); nanometre scale elemental maps and crystal defect characterisation (dislocations, grain boundaries, voids, stacking faults). The ultimate lateral resolution is about 0.2 nm. The strength of this method is in the highest spatial resolution for elemental mapping, small area crystallographic information and strong contrast between crystalline and amorphous phases. However, it requires significant time for sample preparation (several hours), small

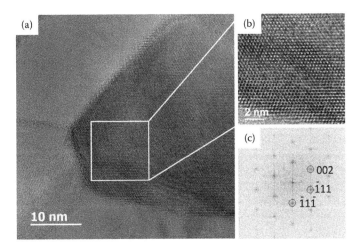

FIGURE 1.17   HR-TEM image (a, b) and fast Fourier transform of selected area (c) of ZrO_2 cubic-nanocrystal with grain boundaries.

sampling volumes and no possibility to observe samples that are unstable under the high-energy electron beam. The utility of TEM methods will be demonstrated in the next chapters.

1.5 APPLICATIONS

The composites and hybrid systems offer various applications. Using these building blocks, the devices for medical application, biological labelling, gas sensors, light-emitting diodes, elements for optical communication, alternative energy sources, memory devices, etc. can be elaborated and produced.

The development of these materials also stimulates the progress in fundamental physics and materials science. New theoretical concepts arise, allowing us to go deeper into the understanding of different novel physical and chemical phenomena appearing in the composites, with different properties from their parent materials. This helps also in the development of non-destructive express methods for materials characterisation, in the elaboration of different approaches and receipts for the monitoring of the parameters of the devices aiming at improving operating stability and minimising degradation phenomena. More detailed application of composites and hybrid systems addressed in this book can be found in corresponding chapters.

REFERENCES

Agocs, E., Zolnai, Z., Rossall, A. K., van den Berg, J. A., Fodor, B., Lehninger, D., Khomenkova, L., Ponomaryov, Gudymenko, O., Yukhymchuk, V., Kalas, B., Heitmann, J. and Petrik, P. 2017. Optical and structural characterization of Ge clusters embedded in ZrO_2. *Appl Surf Sci* 421:283–8.

Athauda, T. J., Madduma-Bandarage, U. S. K. and Vasquez, Y. 2014. Integration of ZnO/ZnS nanostructured materials into a cotton fabric platform. *RSC Adv* 4:61327.

Bang, J. H. and Suslick, K. 2010. Applications of ultrasound to the synthesis of nanostructured materials. *Adv Mater* 22:1039–59.

Barile, C. J., Slotcavage, D. J., Jingye Hou, Strand, M. T., Hernandez, T. S. and McGehee, M. D. 2017. Dynamic windows with neutral color, high contrast, and excellent durability using reversible metal electrodeposition. *Joule* 1:133–45.

Camargo, P. H. C., Satyanarayana, K. G. and Wypych, F. 2009. Nanocomposites: Synthesis, structure, properties and new application opportunities. *Mat Res* 12:1–39.

Chan, K. Y., Voskanyan, A. and Li, C. Y. V. 2017. Method of producing a porous crystalline material with a highly uniform structure. Patent WO 2017/143978 A1, Pub. Date: 2017-08-31.

Chen, B., Meng, Y., Sha, J., Zhong, C., Hu, W. and Zhao, N. 2018. Preparation of MoS_2/TiO_2 based nanocomposites for photocatalysis and rechargeable batteries: Progress, challenges, and perspective. *Nanoscale* 10:34–68.

Cobianu, C. P., Dumitru, V. G., Serban, B.-C., Stratulat, A., Brezeanu, M. and Buiu, O. 2016. Metal oxide nanocomposite heterostructure methods and hydrogen sulfide sensors including the same. US Patent, Pub. No. 2016/0011161 A1, Pub. Date: 2016-01-14. Washington, DC: U.S. Patent and Trademark Office.

Cobianu, C. P., Dumitru, V. G., Serban, B.-C., Stratulat, A., Brezeanu, M. and Buiu, O. 2017. Fet and fiber based sensor. US Patent, Pub. No. 2017/0067849 A1. Pub. Date: 2017-03-09. Washington, DC: U.S. Patent and Trademark Office.

Cueff, S., Labbé, C., Jambois, O., Garrido, B., Portier, X. and Rizk, R. 2011. Thickness-dependent optimization of Er^{3+} light emission from silicon-rich silicon oxide thin films. *Nanoscale Res Lett* 6:395.

Debieu, O., Pratibha Nalini, R., Cardin, J., Portier, X., Perrière, J. and Gourbilleau, F. 2013. Structural and optical characterization of pure Si-rich nitride thin films. *Nanoscale Res Lett* 8:31.

Doğan, I. and van de Sanden, M. C. M. 2013. *Direct characterization of nanocrystal size distribution using Raman spectroscopy. J Appl Phys* 114:134310.

Gedanken, A., Nitzan, Y., Perelshtein, I., Perkas, N. and Applerot, G. 2016. Sonochemical coating of textiles with metal oxide nanoparticles for antimicrobial fabrics. US Patent, Pub. No. 2016/0302420 A1. Pub. Date: 2016-10-20. Washington, DC: U.S. Patent and Trademark Office.

George, T., Deshpande, V. S., Sharp, K. and Wadley, H. N. G. 2014. Hybrid core carbon fiber composite sandwich panels: Fabrication and mechanical response. *Composite Structures* 108:696–710.

Granqvist, C. G. 2014. Electrochromics for smart windows: Oxide-based thin films and devices. *Thin Solid Films* 564:1–38.

Gourbilleau, F., Ternon, C., Maestre, D., Palais, O. and Dufour C. 2009. Silicon-rich SiO_2/SiO_2 multilayers: A promising material for the third generation of solar cell. *J Appl Phys* 106:013501.

Guo, I. W., Pekcevik, I. C., Wang, M. C. P., Pilapil, B. K. and Gates, B. D. 2014. Colloidal core-shell materials with 'spiky' surfaces assembled from gold nanorods. *Chem Commun* 50: 8157–60.

Heinz, H., Pramanik, C., Heinz, O., Ding, Y., Mishra, R. K., Marchon, D., Flatt, R. J., Estrela-Lopis, I., Llop, J., Moya, S. and Ziolo, R. F. 2017. Nanoparticle decoration with surfactants: Molecular interactions, assembly, and applications. *Surf Sci Rep* 72:1–58.

Kao, J., Thorkelsson, K., Bai, P., Rancatore, B. J. and Xu, T. 2013. Toward functional nanocomposites: Taking the best of nanoparticles, polymers, and small molecules. *Chem Soc Rev* 42:2654–78.

Kim, Y., Lee, J., Yeom, M. S., Shin, J. W., Kim, H., Cui, Y., Kysar, J. W., Hone, J., Jung, Y., Jeon, S. and Han, S. M. 2013. Strengthening effect of single-atomic-layer graphene in metal-graphene nanolayered composites. *Nat Commun* 4:2114.

Khomenkova, L., Labbé, C., Portier, X., Carrada, M. and Gourbilleau, F. 2013. Undoped and Nd^{3+} doped Si-based single layers and superlattices for photonic applications. *Phys Stat Sol A* 210:1532–1543.

Kumar, A., Yadav, N., Bhatt, M., Mishra, N. K., Chaudhary, P. and Singh, R. 2015. Sol-gel derived nanomaterials and it's applications: A review. *Res J Chem Sci* 5:98–105.

Lee, J.-S., Kovalenko, M. V., Huang, J., Chung, D. S. and Talapin, D. V. 2011. Band-like transport, high electron mobility and high photoconductivity in all-inorganic nanocrystal arrays. *Nature Nanotechnol* 6:348–52.

Lee, S. W., Cheon, S. A., Kim, M. I. and Park, T. J. 2015. Organic-inorganic hybrid nanoflowers: types, characteristics, and future prospects. *J Nanobiotechnol* 13:54.

Liang, C.-H., Debieu, O., An, Y.-T., Khomenkova, L., Cardin, J. and Gourbilleau, F. 2012. Effect of the Si excess on the structure and the optical properties of Nd-doped Si-rich silicon oxide. *J Lumin* 132:3118–21.

Magdziarz, A. and Colmenares, J. C. 2017. In situ coupling of ultrasound to electro-and photo-deposition methods for materials synthesis. *Molecules* 22:216.

Martina, K., Tagliapietra, S., Barge, A. and Cravotto, G. 2016. Combined microwaves/ultrasound, a hybrid technology. *Top Curr Chem (Z)* 374:79.

Nag, A. and Talapin, D. V. 2014. Materials and methods for the preparation of nanocomposites. US Patent, Pub. No. 2014/0346442 A1, Pub. Date: 2014-11-27. Washington, DC: U.S. Patent and Trademark Office.

Otero, R., Vázquez de Parga, A. L. and Gallego, J. M. 2017. Electronic, structural and chemical effects of charge-transfer at organic/inorganic interfaces. *Surf Sci Rep* 72:105–45.

Owens, G. J., Singh, R. K., Foroutan F., Alqaysi, M., Han, C.-H., Mahapatra, C., Kim, H.-W. and Knowles, J. 2016. Sol-gel based materials for biomedical applications. *Prog Mater Sci* 7:1–79.

Pandey, S. and Mishra, S. B. 2011. Sol-gel derived organic-inorganic hybrid materials: Synthesis, characterizations and applications. *J Sol-Gel Sci Technol* 59:73–94.

Ponomaryov, S. S., Yukhymchuk, V. O., Lytvyn, P. M. and Valakh M. Ya. 2016. Direct determination of 3D distribution of elemental composition in single semiconductor nanoislands by scanning auger microscopy. *Nanoscale Res Lett* 11:103.

Santhanam, K. S. V., Kandlikar, S. G, Mejia, V. and Yang Yue. 2016. Electrochemical process for producing graphene, graphene oxide, metal composites and coated substrates. US Patent, Pub. No. 2016/0017502 A1, Pub. Date: 2016-01-21. Washington, DC: U.S. Patent and Trademark Office.

Shvalagin, V. V., Stroyuk A. L., Kotenko, I. E. and Kuchmii, S. Ya. 2007. Photocatalytic formation of porous CdS/ZnO nanospheres and CdS nanotubes. *Theor Experim Chem*, 43:215–19.

Stroyuk, A. L., Shvalagin, V. V., Raevskaya, A. E., Kryukov, A. I. and Kuchmii, S. Ya. 2008. Photochemical formation of semiconducting nanostructures. *Theor Experim Chem* 44:205–231.

Sun, J., Ge, J., Liu, W., Lan, M., Zhang, H., Wang, P. and Niu, Z. 2014. Multi-enzyme co-embedded organic-inorganic hybrid nanoflowers: synthesis and application as a colorimetric sensor. *Nanoscale* 6:255–62.

Tada, H., Mitsui, T., Kiyonaga, T., Akita, T. and Tanaka, K. 2006. All-solid-state Z-scheme in CdS-Au-TiO$_2$ three-component nanojunction system. *Nature Mater* 5:782–6.

Tuantranont, A., Sriprachabwong, C., Wisitsoraat, A. and Phokharatkul, D. 2012. A method for preparing polymer/oxygen-free graphene composites using electrochemical process. WO Patent, Pub. No. 2012064292 A1. Pub. Date: 2012-05-18.

Vukomanovic, M., Skapin, S. D. and Suvorov, D. 2015. Functionalized hydroxyapatite/gold composites as "green" materials with antibacterial activity and the process for preparing and use thereof. EP Patent, Pub. No. 2863751 A1. Pub. Date: 2015-04-29. European Patent Office.

Xu, H., Zeiger, B. W. and Suslick, K. S. 2013. Sonochemical synthesis of nanomaterials. *Chem Soc Rev* 42: 2555–67.

CHAPTER 2

Multifunctional Zirconia-based Nanocomposites

Nadiia Korsunska and Larysa Khomenkova

CONTENTS

2.1 INTRODUCTION

Zirconia (ZrO_2) is a multifunctional material which can be found in the different forms: crystals, ceramics, films and powders, including nanopowders. Due to unique mechanical, electrical, thermal and optical properties, it has diverse applications as high temperature and corrosion resistant coatings (Schulz et al. 1996; Padture et al. 2002; Habibi et al. 2013), radiation detectors (Kirm et al. 2005), biological labelling (Wang et al. 2010), catalysts (Sun et al. 1994), oxygen sensors (Izu et al. 2009; Fidelius et al. 2009, Radhakrishnan et al. 2012; You et al. 2017), solid-oxide fuel cells (SOFCs) (Riley 1990; Mahato et al. 2015; Kumar et al. 2017; Liu et al. 2017), etc. However, for a set of important applications, not only pure ZrO_2, but also the composites prepared by the addition of other metal oxides, are often in demand. The incorporation of complementary cations in the crystal lattice of zirconia affects its properties, first of all, its crystal structure. Namely, the latter is often a crucial factor in many applications.

At room temperature, ZrO_2 adopts a monoclinic crystal structure. Its transformation into tetragonal (*t*) and cubic (*c*) phases occurs at the temperatures exceeding essentially 1000°C. Moreover, the temperature of phase transition depends on ZrO_2 crystallite sizes. However, both of these phases are often required at lower temperatures.

One of the approaches used for low-temperature stabilisation of the *t*- and *c*-phases is a doping of ZrO_2 with subvalent cations (as Y, Ca, Sc, etc.). Yttrium is the most commonly used additive, and it can be retained in ZrO_2 at higher temperatures (Zhang et al. 2007; Sun et al. 2007; Restivo et al. 2009; Korsunska et al. 2014a; Asadikiya et al. 2018).

The main factor of the stabilisation of the *t*- and *c*-phases is the formation of oxygen vacancies which promote the oxygen diffusion through the crystal structure and control the ionic conductivity of zirconia. The latter effect is the basis of the operation of the electrolyte in SOFC and oxygen sensors (Fidelus et al. 2009; Kajiyama and Nakamura 2015; Reiter and Seyr 2016; Fischer et al. 2017) and SOFCs (Zhang et al. 2007; Sun et al. 2007; Restivo et al. 2009; Asadikiya et al. 2018). On the other hand, the tetragonal phase is desirable for catalyst preparation because it usually shows the better catalytic properties (Ma et al. 2005; Wang et al. 2007; Pakharukova et al. 2015).

In some cases, the doping with more than one additive can be necessary. For instance, to improve catalytic activity of Y-stabilised ZrO_2 (YSZ), such composites are also doped with Cu, Pt or Ag (Pakharukova et al. 2015; Gorban et al. 2017; Singhania and Gupta 2017). Another example is ZrO_2-based electrolyte for SOFCs. As compared to YSZ composites, ZrO_2 co-doped with Ce and Sc showed higher ionic conductivity (Hearing et al. 2005; Lee et al. 2005; Vasylyev et al. 2018).

At the same time, the YSZ composite, being thermally stable, has low electronic conductivity. When aiming for its application as a SOFC anode, high electronic conductivity should be achieved. For this purpose, an addition of metallic Ni- or Cu-powders was proposed. A successful application of such alloys as anode materials was demonstrated (Minh 1993; Sun et al. 2007; Restivo et al. 2009; Vasylyev et al. 2018).

It is worth pointing out that subvalent impurities, being incorporated into the ZrO_2 grains, affect not only their structure but also other characteristics of the composites. In particular, the zirconia doping with rare-earth elements results in the appearance of an intense luminescence (Smits et al. 2014, 2017). A specific green band caused by Cu inside the grains was also observed in photo- and cathodoluminescence spectra of (Cu, Y) co-doped ZrO_2 composite (Korsunska et al. 2015a, 2015b, 2018). At the same time, copper localised on the grain surface or in its near-surface region stipulates such a composite application as catalyst (Sun et al. 1994; Zhang et al. 2014; Samson et al. 2014; Pakharukova et al. 2015). It also determines fungicidal properties of Cu-doped ZrO_2 powders and their tribological behaviour, as well as demonstrates the ability to compact the ceramic (Sherif et al. 1980; Ran et al. 2007; Kong et al. 2014). In this case, the nature of surface Cu species and their surface concentration as well as surface area play an important role.

To obtain the required properties, different techniques for composite preparation are used. Both the crystalline structure and the kind of Cu-loaded substances on the grain surface can be monitored by the variation of preparation approach. For instance, for SOFC anode preparation, the mechanical alloying of the YSZ and metallic Cu-powders is applied. In this case, the formation of metallic Cu in the YSZ pores (i.e. on the grain surface) provides good electronic conductivity (Sun et al. 2007; Gross et al. 2007; Restivo et al. 2009). At the same time, the impregnation of ZrO_2 or YSZ powders with copper nitrates is usually used for catalyst preparation (Samson et al. 2014; Pakharukova et al. 2015) and results mainly in the enrichment with copper of the grain surface or near-surface region, while the co-precipitation technique allows obtaining simultaneous Cu presence in the grain bulk and on the grain surface (Korsunska et al. 2017b, 2017c).

The co-doping of ZrO_2 with several dopants can change the composite properties as compared to the singly doped ZrO_2 (crystal structure, luminescence intensity, electrical properties, etc.). In fact, as shown by Winnubst et al. (2009), the Cu addition to tetragonal YSZ, prepared by the co-precipitation technique, leads to the decrease of the temperature of tetragonal-to-monoclinic phase transformation. The similar effect is also caused by the incorporation of Nb in rare-earth-stabilised ZrO_2 tetragonal phase (Smits et al. 2017).

In this chapter, some recently observed effects of co-doping as well as the influence of technological conditions (temperature of annealing, cooling rate, doping level, etc.) on structure, luminescence and surface properties of composites and dopant relocation are described.

2.2 STRUCTURAL PROPERTIES OF PURE AND DOPED ZIRCONIA

As mentioned above, at room temperature zirconia has a monoclinic (*m*) crystal structure, whereas tetragonal (*t*) or cubic (*c*) phases under equilibrium conditions are stable at higher temperatures. Normally, as the phase diagram shows (Figure 2.1), the high temperature *t*-phase in pure unconstrained large crystals is stable only above 1205°C (1478 K), while the cubic one is stable above 2377°C (2650 K) (Massalski et al. 1990).

However, the temperature of phase transformation decreases with the decrease of crystallite sizes (Garvie and Goss 1986). In the case of nanopowders, the tetragonal or cubic phase structure, depending on crystallite size, can be stable even at room temperature. Specifically, at smaller size, ZrO_2 grains have a cubic structure at room temperature, while size increase results in phase transformation to monoclinic one (*c-m* transformation). For example, in ZrO_2 nanopowders synthesised via reactive plasma processing, the crystallites smaller than 6 nm are of cubic phase; the tetragonal phase is found to be stable for crystallites of smaller than 20 nm, while the particles of larger size have the monoclinic structure (Jayakumar et al. 2013). One of the reasons for tetragonal phase stabilisation is excess surface energy (Garvie and Goss 1986). If the particle size is reduced, the contribution of surface energy to the total energy increases, resulting in *t*-phase stabilisation (Tok et al. 2006).

On the other hand, it was shown that stabilisation of the cubic phase at room temperature in nanocrystalline ZrO_2 is caused by the formation of oxygen vacancies due to local disorder and distortion of the oxygen sublattice (Soo et al. 2008) that is in accordance with earlier theoretical studies (S. Fabris et al. 2002). At the same time, in the case of ZrO_2 films, the mechanical stresses between film and substrate are also considered as a factor stabilising the cubic phase (Pugachevskii et al. 2011). As it was mentioned in Section 2.1,

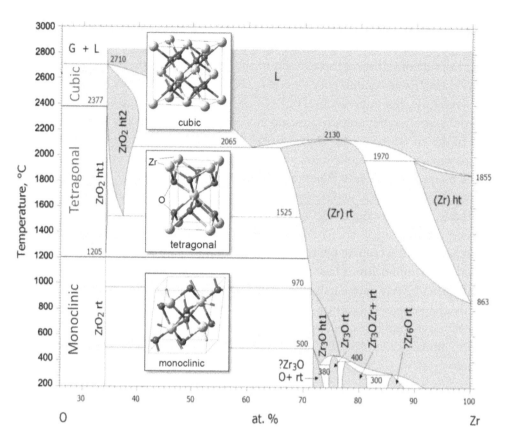

FIGURE 2.1 O-Zr phase diagram and structures of ZrO_2 in the room temperature (rt) monoclinic phase, the high temperature (ht1) tetragonal phase and the high temperature (ht2) cubic phase. The phase transitions are observed at 1205°C (1478 K) and 2377°C (2650 K). (The phase diagram is adapted from Massalski, T. B., et al., *Binary Alloy Phase Diagrams*, ASM International, Materials Park, OH, 1990.)

to stabilise the tetragonal or cubic zirconia phases at room temperature, the doping with different subvalent metallic atoms which stimulate oxygen vacancies formation is usually used. In this case, the main factor of their formation is the requirement for the charge compensation of these impurities (acceptors) by vacancies (donors). The stabilisation of the tetragonal or cubic phase is supposed to be due to lattice relaxation caused by oxygen vacancies, and one of the stabilisation factors is considered to be the increase of Zr-O covalent bonding (Yuan et al. 1999). At the same time, the supervalent cation impurities due to their donor nature can result in the decrease of the content of oxygen vacancy stimulated by the doping with subvalent impurities (Smits et al. 2014, 2017). Thus, the co-doping with different dopants can considerably transform the nanocrystal structure.

Since simultaneous doping with different dopants is frequently used in zirconia-based composites intended for different application (as SOFC electrolyte, catalysts, the powders for light-emitting devices, especially doped with rare-earth elements, etc.), this aspect will be considered in more detail in the next sections, along with the examples of different phenomena stimulated by co-doping.

2.3 (Sc, Ce)-co-doped ZrO₂ AS AN ELECTROLYTE FOR SOLID OXIDE FUEL CELLS

Currently SOFCs based on ZrO_2-related composites are of great interest as alternative energy sources (energy conversion systems) due to their higher energy efficiency and environmental friendliness.

The maximum ionic conductivity in ZrO_2-based systems is observed when the concentration of acceptor-type dopant(s) is close to the minimum necessary to completely stabilise the cubic phase (Etsell and Flengas 1970; Kharton et al. 1999; Yamamoto et al. 1998; Badwal 1992). This concentration depends, to some extent, on the preparation procedure (Kharton et al. 2004). Nevertheless, for most important systems, the dopant concentration ranges providing maximum ionic conductivity are well known.

Currently, Y is the most widely applied dopant in commercial SOFC electrolytes (Luo et al. 2000). The maximum ionic conductivity has been found for 8 mol% of Y_2O_3 (Kilo et al. 2003) (Figure 2.2). Further additions decrease the ionic conductivity due to increasing association of the oxygen vacancies and dopant cations into complex defects of low mobility. This tendency increases with increasing difference between the host and dopant cation radii (Inaba and Tagawa 1996; Bouwmeester and Burggraaf 1996; Sammes et al. 1999; Mogensen et al. 2000; Yamamoto et al. 1998).

According to these findings, the high ionic conductivity of the YSZ with 8 mol% Y_2O_3 composite is realised at high temperatures (the conductivity is 0.1 S/cm at 1000°C and 0.03 S/cm at 800°C). Therefore, this composite requires high enough operating temperatures (nearly 1000°C) that cause the degradation of SOFC's component after a long operating time (Hong et al. 2015). To reduce the operating temperatures, the thinning of electrolytes was proposed, but this approach is restricted by the mechanical strength of fuel cells. For instance, anode-supported SOFCs could somewhat reduce the operating temperature down to ~800°C. However, when the thickness decreases to ~10 μm, an additional

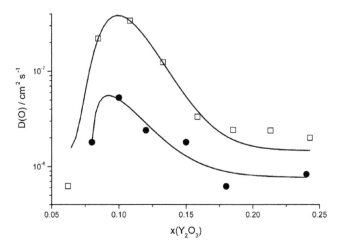

FIGURE 2.2 Diffusion coefficients in Y-stabilised ZrO_2 at 700°C (973 K) as a function of the Y content. (Reproduced with permission from Kilo et al. (2003). Copyright 2003, The Royal Chemical Society.)

reduction in thickness does not result in the noticeable further reducing of resistance and promotes the defect formation in excessively thin electrolytes. A more effective approach consists in the increase of ionic conductivity of electrolytes. For this purpose, other impurities were tested for zirconia-based composites. The temperature dependencies of oxygen conductivity of different electrolytes are shown in Figure 2.3.

According to many reports on the conductivity of various zirconia-based electrolytes, ZrO_2 doped with Sc_2O_3 has the highest ionic conductivity (Killner et al. 1982; Riley 1990; Lee et al. 2005; Omar et al. 2011; Abbas et al. 2011), which maximum is observed in $Zr_{1-x}Sc_xO_{2-x/2}$ ceramics at $x = 0.09 - 0.11$ (Figure 2.4).

It is known that among four main phases (monoclinic, tetragonal, rhombohedral β and rhombohedral γ) of Sc-stabilised (ScSZ) systems, the highest ionic conductivity suitable for use in a solid electrolyte for SOFCs has a cubic one. The phase diagram of Sc_2O_3-doped zirconia is shown in Figure 2.5 (Chiba et al. 1997). It is seen that the cubic phase is a main phase when the doping level is in the range of 8–12 mol% Sc_2O_3. However, in this Sc_2O_3 stabilised ZrO_2 system, the transition of the highly conductive cubic phase to rhombohedral around 650°C during cooling or due to ageing at high temperatures (700°C–1000°C) occurs, which results in the sharp reduction of the conductivity (Figure 2.4).

To stabilise the cubic phase of ScSZ materials at low temperatures required for operation of low- or intermediate-temperature SOFCs, various additional dopants like CeO_2, Sm_2O_3, Yb_2O_3 and Al_2O_3 were considered (Haering et al. 2005). Among different co-doped oxides,

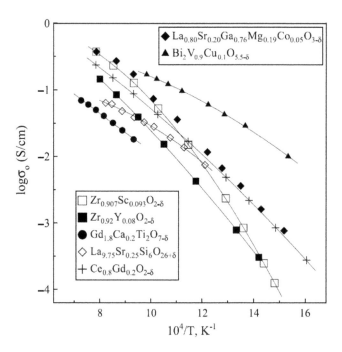

FIGURE 2.3 Temperature dependencies of oxygen conductivity of different electrolytes. (Reproduced with permission from *Solid State Ionics*, 174, Kharton, V. V. et al., Transport properties of solid oxide electrolyte ceramics: A brief review, 135–149, Copyright 2004, Elsevier.)

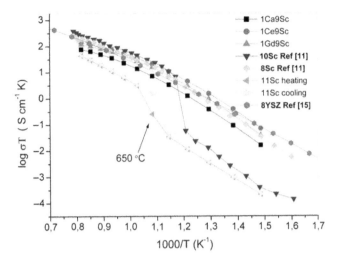

FIGURE 2.4 The Arrhenius plot of the conductivity of co-doped ZrO_2-Sc_2O_3 system. For comparison, data for Ca-, Ce- and Gd-contained composites, as well as YSZ-8, are also included. (Reproduced with permission from *Solid State Ionics*, 184, Abbas, H. A. et al., Preparation and conductivity of ternary scandia-stabilised zirconia, 6–9, Copyright 2011, Elsevier.)

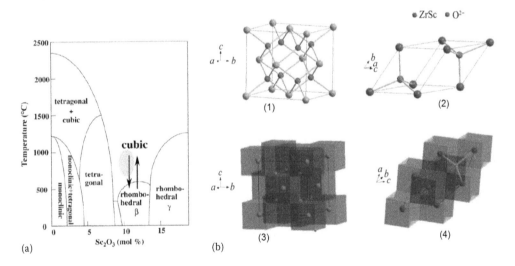

FIGURE 2.5 (a) Phase diagram of Sc_2O_3-ZrO_2 binary system; (b) structure models (1, 2) and structure coordinate-models (3, 4) of cubic (1, 3) and rhombohedral (2, 4) ScSZ.

the highest ionic conductivity was measured for 1 mol% CeO_2 and 10 mol% Sc_2O_3 added to ZrO_2 (i.e. for 10Sc1CeSZ), which was proposed for SOFC application (Haering et al. 2005).

As the investigation of the local atomic structure of cubic-stabilised ZrO_2 shows, the doping with additional cations results in a large distortion in the first-shell oxygen distribution, which is increased with cation radius (Villella et al. 2001). Therefore, the mechanism of the cubic phase stabilisation with Ce doping was proposed to be caused by the formation of eight-fold coordination with oxygen by the larger tetravalent dopant cation, while oxygen vacancies are coupled with the Zr ions (Arachi et al. 2001).

2.4 (Y, Cu)-co-doped ZrO₂ COMPOSITES FOR CATALYSIS: INFLUENCE OF CALCINATION TEMPERATURES ON PHASE TRANSFORMATION, IMPURITY SEGREGATION AND CATALYTIC ACTIVITY OF COMPOSITES

In recent years, Cu-ZrO₂ as well as (Cu, Y)-ZrO₂ composites have been intensively studied, demonstrating the dependence of their properties on the spatial copper localisation (inside crystallites or at their surface). It was mentioned in Section 2.1 that copper, being present in the near-surface area of crystallites or at their surface, is responsible for the catalytic activity of the composite (Sun et al. 1994; Samson et al. 2014; Zhang et al. 2014; Pakharukova et al. 2015) and its antibacterial properties (Sherif et al. 1980), and also contributes to the sintering of the ceramics (Winnubst et al. 2009; Zhang et al. 2008a). At the same time, copper, which is localised inside nanocrystals, affects their structural characteristics (Zhang et al. 2008b; Bhagwat et al. 2003). The increase of Cu content favoured the decrease of crystalline size and stabilisation of tetragonal and cubic polymorphs (Asadi et al. 2012). Besides, the YZS doping with Cu stimulates the decrease of the temperature of transformation of the tetragonal phase, which at 3% Y is stable until 1000°C, to monoclinic one (Winnubst et al. 2009). However, it was found that their simultaneous attendance in zirconia changes the composite structure in more complex ways.

Below, the effect of Y- and Cu-co-doping on the structure of ZrO₂ as well as the possible mechanism of phase transformation and the impurity segregation will be considered for ZrO₂ nanopowders with 3 mol% Y using the results obtained with different investigation methods. The samples were synthesised by a co-precipitation technique using Zr, Y and Cu nitrates with CuO concentration of 1 mol% (Cu-1) or 8 mol% (Cu-8) (called Cu-1 or Cu-8 samples) and were calcined at different temperatures $T_c = 500°C–1100°C$ for 2 hours and cooled slowly with the furnace. Besides, the Cu-8 sample calcined at 1100°C and rapidly cooled to room temperature was also studied.

2.4.1 XRD Data

XRD patterns of Cu-1 and Cu-8 samples annealed at 500°C–1100°C are shown in Figure 2.6a–d. Because the proximity of the most positions of XRD peaks corresponded to tetragonal and cubic phases, the peaks in the range of 73°–75° were used to separate their contribution. It is known that the cubic phase of pure ZrO₂ shows in this range only a single peak at 74.33, while the tetragonal phase has two peaks at 73.047° and 74.579°. It should be noted that for the samples investigated, the XRD peak positions somewhat differ from those mentioned above due to the doping of ZrO₂ with yttrium and copper. In spite of this, the comparison of peak position and its number in the range 73°–75° for XRD patterns of different samples allows discrimination of both t- and c-phases.

As can be seen in Figure 2.6, at low calcination temperatures (500°C–600°C) there is only one peak in the range of 73°–75°, which may be caused by the dominant contribution of the cubic phase or by the small size of t-phase nanocrystals, which results in the overlapping of the two peaks. However, the asymmetric shape of this peak as well as the proximity of its position to high-energy peaks in the samples calcined at higher temperatures (and clearly showing two peaks, which is characteristic of t-phase) testifies to significant contribution of tetragonal phase. This statement is confirmed by the Raman scattering spectra

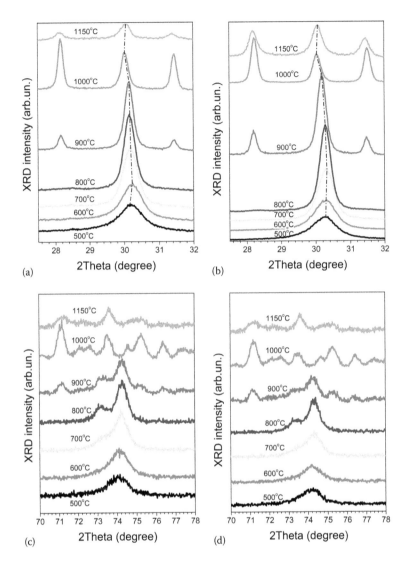

FIGURE 2.6   XRD patterns of Cu-1 (a, b) and Cu-8 (c, d) samples calcined at 500°C–1100°C.

(Korsunska et al. 2015b) and the Rietveld refinement of XRD patterns. The latter shows the presence of a very low content of cubic phase (1%–3%) (Korsunska et al. 2017a).

The tetragonal phase is clearly observed for all samples calcined at 700°C–900°C and until 900°C this phase is dominant, but at $T_c = 1000$°C the transformation of the tetragonal phase into the cubic one occurs. This phase is also present at $T_c = 1100$°C. At $T_c > 700$°C, in addition to t- or c- phases, the monoclinic one appears. Besides, in Cu-8 samples calcined at $T_c > 900$°C–1100°C, XRD peaks at 35.5°, 38.8° and 68°, are detected, caused by the presence of CuO (Korsunska et al. 2017c).

The estimation of coherent domain sizes of tetragonal (cubic) and monoclinic phases shows that the crystallite mean size for tetragonal (cubic) phase increases gradually from ~11.4 nm (500°C) up to 35–39 nm (1000°C) for both sets of samples. For the monoclinic phase, this value is in the interval 29–51 nm.

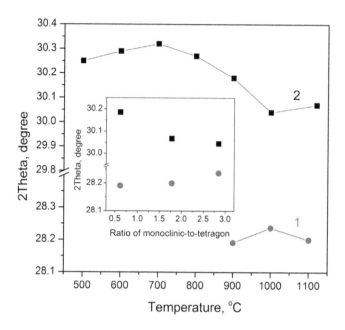

FIGURE 2.7 The dependencies of peak positions of the monoclinic phase (at ~28°) (curve 1) and tetragonal phase (at ~30°) (curve 2) on T_c for Cu-8 samples. Inset shows these peak positions versus monoclinic-to-tetragonal phase ratio.

The investigation of the shift of XRD peak position with calcination temperatures gives additional information about the phase transformation and the processes of impurity segregation. It was found that all reflections of Cu-1 and Cu-8 samples are shifted to higher angles in comparison with the corresponding peak positions for ZrO_2 samples doped with 3 mol% Y_2O_3. Essentially, this shift is higher in Cu-8 samples, testifying to Cu incorporation in the grains. The temperature dependence of peak position at ~30°, which corresponds to the tetragonal (or cubic) phase and is not overlapped with the reflections of the monoclinic one, is shown in Figure 2.7 for Cu-8 samples (curve 2).

As Figure 2.7 shows, this dependence is non-monotonic: when T_c increases, this peak at first shifts to higher angles (up to 700°C), then (at $T_c = 700°C–800°C$) it moves to lower ones and finally (at 1100°C) shifts to higher angles again. It is interesting that for Cu-1 and Cu-8 samples calcined at 900°C, the XRD peaks are close to each other and approach the peak position observed in ZrO_2-3%Y_2O_3 (PDF#01-071-4810). However, the sharpest shift to lower angles takes place at 1000°C, when the contribution of monoclinic phase is the highest. At the same time, the shift of m-phase reflections with T_c increase occurs to the opposite side as compared to the tetragonal peaks, and its value depends on m/t phase relation (Figure 2.7, inset).

2.4.2 TEM Observation

TEM study was performed for both sets of samples. However, hereafter, only the results for Cu-8 samples are presented because they demonstrate more pronounced variations of their XRD data with T_c. Figure 2.8 shows the evolution of the grain structure with T_c. It is seen that the xerogel sample has an amorphous structure that is in agreement with XRD data.

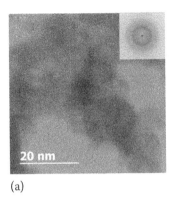 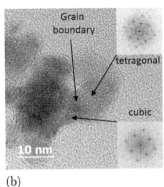 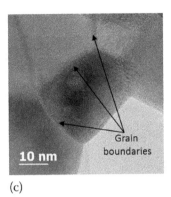

(a)   (b)   (c)

FIGURE 2.8   High-resolution TEM images of Cu-8 samples: initial xerogel (a) and after calcination at $T_c = 600°C$ (b) and 900°C (c). (Reproduced from *Nanoscale Res Lett*, Impurity-governed modification of optical and structural properties of ZrO_2-based composites doped with Cu and Y, 12, 2017a, 157, Korsunska, N. et al., Figure 3, Copyright 2017 Nanoscale Research Letters, Springer, under the Creative Commons Attribution 4.0 International License (http://creativecommons.org/licenses/by/4.0/).)

The T_c increase results in the formation of nanocrystals. Their mean size increases from ~14 nm ($T_c = 600°C$) to ~46 nm ($T_c = 900°C$). For $T_c = 600°C$, the nanocrystals were found to be with tetragonal and cubic structures (Figure 2.8, b), while for $T_c = 900°C$, the main part of nanocrystals was of the monoclinic phase (Figure 2.8, c).

Besides, large particles consisting of smaller nanocrystals with grain boundaries were found. The presence of nanocrystals with tetragonal and cubic structure in Cu-8 samples calcined at $T_c = 900°C$ was also detected, but their amount was low.

More interestingly, for higher T_c, the segregations of Y and Cu at grain boundaries were observed. In fact, the STEM-EDX analysis of a grain boundary (line-scan along the line in Figure 2.9a) for the Cu-8 sample calcined at 900°C shows that these grain boundaries are enriched in Y and Cu (Figure 2.9b,c). It is worth noting that the Cu signal is higher than that of Zr despite the low content of Cu compared to that of Zr because of the use of a holey carbon-covered copper grid. Besides Y and Cu segregation, the formation of the dark circle-like regions was observed within the grains (Figure 2.9a). They correspond to the cavities that are depleted for all elements. One of the reasons for their formation is the segregation of vacancies upon annealing (Korsunska et al. 2017a).

Thus, TEM study shows that the T_c increase favours the formation of nanocrystals and their sintering at high T_c as well as the segregations of Y and Cu at grain boundaries.

2.4.3 Diffuse Reflectance Spectra

The effect of Cu segregation at the grain surface was confirmed also by diffuse reflectance spectra. Besides, they give the information on oxygen vacancies content.

Diffuse reflectance (DR) spectra of Cu-1 and Cu-8 samples calcined at different temperatures are shown in Figure 2.10a,b. These DR spectra contain the absorption band (peaked at ~270 nm) near the band edge of ZrO_2, which is more intensive for Cu-8 samples. An increase of T_c in the 500°C–700°C range leads to a slight increase of its intensity in

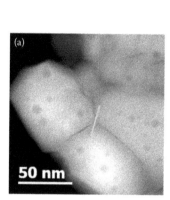
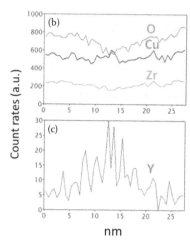

FIGURE 2.9 (a) STEM HAADF image of Cu-8 powder calcined at 900°C. The dark regions with circular shapes in the grains correspond to the cavities. (b, c) EDX profiles across the grain boundary for O, Cu and Zr (b) and Y (c) elements. (Reproduced from *Nanoscale Res Lett*, Impurity-governed modification of optical and structural properties of ZrO_2-based composites doped with Cu and Y, 12, 2017a, 157, Korsunska, N. et al., Figure 4, Copyright Nanoscale Research Letters, Springer, Copyright 2017 Nanoscale Research Letters, Springer, under the Creative Commons Attribution 4.0 International License (http://creativecommons.org/licenses/by/4.0/).)

Cu-8 samples, while it is almost unchanged for Cu-1 samples. For $T_c = 800°C–1000°C$, the intensity of this band decreases essentially in both groups of samples.

In addition to the 270-nm absorption band, another one is observed in the range of 600–900 nm (Figure 2.10). Its intensity increases with T_c rising up to 800°C, being accompanied by the shift of its peak position to shorter wavelengths. For higher calcination temperatures, the intensity of this band decreases and the edge of the fundamental absorption of crystalline CuO appears for Cu-1 and Cu-8 samples (Figure 2.10a,b).

The band in the 600–900 nm range is usually attributed to d-d transitions of the Cu^{2+} ions in an octahedral or tetragonal distorted octahedral surrounding and associated with dispersed CuO on the surface of the nanocrystals, or with Cu_{Zr} substitutional atoms located in the near-surface region (Goff et al. 1999; Pakharukova et al. 2009, Samson et al. 2014). The increase of Cu loading caused the increase of intensity and the short-wavelength shift of this Cu-related band. The latter was attributed to an increase of octahedral distortion (Pakharukova et al. 2009). It should be noted that intensity of the Cu-related band increases, and dispersed Cu clusters give the contribution to DR spectra in visible spectral range due to localised surface plasmon resonance when calcination is accompanied by quenching (Korsunska et al. 2017b).

Based on the obtained experimental data, the nature of the absorption band at ~270 nm can be considered. A similar band was observed in works devoted to the study of monoclinic and tetragonal ZrO_2 doped with Cu by impregnation. The intensity of this band increased with Cu loading, and it was ascribed to electron transitions from oxygen to copper.

However, as can be seen from Figure 2.10a (inset), the same absorption band is present in the ZrO_2 samples doped with yttrium only, and its intensity increases with Y content. The common feature of both Y- and Cu-doped ZrO_2 is the formation of oxygen vacancies

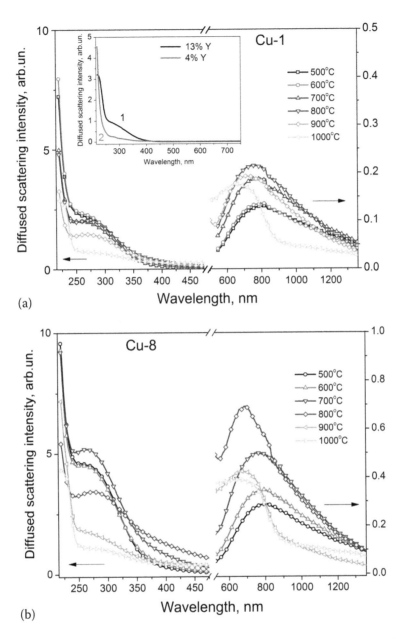

FIGURE 2.10 Diffuse reflectance spectra of Cu-1 (a) and Cu-8 samples (b), calcined at $T_c = 500$–1000°C, as well as the spectra of Cu-free ZrO_2 samples doped with 13% of yttrium (curve 1) and 4% of yttrium (curve 2), $T_c = 700°C$ (inset in (a)).

(Pakharukova et al. 2009). It allows assigning the band at ~270 nm, observed in samples investigated, to oxygen vacancies. Indeed, the intensity of this band decreases at $T_c \geq 800°C$ (Figure 2.10), which correlates with the appearance and increase of the contribution of the monoclinic phase (Figure 2.6), containing, as known, fewer oxygen vacancies than Y- or Cu-stabilised tetragonal phases. Thus, the diffuse reflectance spectra allow the detection of the oxygen vacancies introduced by doping.

It should be noted that in Cu-ZrO$_2$ samples doped by impregnation, the Cu atoms can penetrate into the near-surface region of ZrO$_2$ nanocrystals and even form a solid solution (Pakharukova et al. 2015), which also should create oxygen vacancies. Therefore, in that case, the oxygen vacancies can also contribute to an absorption band at ~270 nm.

The contribution of Cu-surface substances is more pronounced for the samples calcined at higher T_c. One of them is CuO, whose formation was revealed by diffuse reflectance spectra and detected by the XRD method (Korsunska et al. 2017a, 2017b).

2.4.4 EPR Spectra

EPR spectra of studied samples are shown in Figure 2.11. Spectra of Cu-1 and Cu-8 samples annealed at the same temperature are similar. Depending on T_c, there are two types of EPR spectra in our samples. The first type (denoted below as spectrum I) is observed in the samples calcined at $T_c = 500°C–800°C$. Another EPR spectrum (denoted below as spectrum II) containing a set of irregular shape lines in a wide range of magnetic fields is detected in the samples calcined at 800°C–1000°C (Figure 2.11).

The integral intensity of spectrum I for Cu-1 and Cu-8 samples is nearly the same. Analysis of EPR spectra of different samples shows that spectrum I consists of at least three signals, whose intensities depend on T_c. The first component (s1) exhibits the characteristic copper hyperfine splitting and can be described by spin-Hamiltonian parameters $g_\perp = 2.072$, $g_\parallel = 2.32$, $A_\perp \sim 0$ G and $A_\parallel \sim 150$ G. Two others (s2 and s3) are single featureless lines with $g \sim 2.20$ and $g \sim 2.15$, respectively (Korsunska et al. 2017a). Because these signals are absent in Cu-free samples, it can be assumed that they are also caused by Cu-related centres. In this case, the absence of Cu-related hyperfine structure can be explained by the exchange interaction between copper ions. With T_c increase, the intensities of all signals of the spectrum I decrease monotonically (Figure 2.12).

Similar EPR spectra were also observed in other copper-doped oxides (TiO$_2$, ZnO, etc.) and were attributed to copper ions in the Cu-related surface complexes. Specifically,

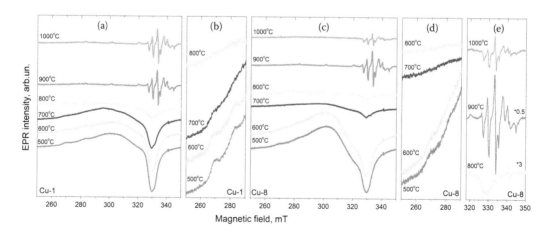

FIGURE 2.11 EPR spectra of Cu-1 (a, b) and Cu-8 (c–e) samples calcined at 500°C–1000°C. The spectra of the samples calcined at $T_c = 500°C–800°C$ are denoted in the text as "spectrum I" type and others as "spectrum II" type.

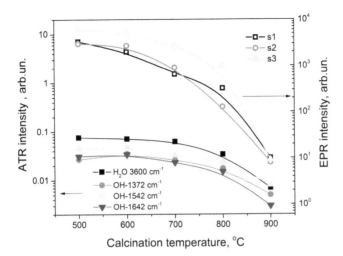

FIGURE 2.12   Variation of the intensities of the components of spectrum I (right y-axis) and ATR intensities of infrared absorption bands of water molecules and OH groups (left y-axis) on the calcination temperature for the Cu-8 samples.

the signal with spin-Hamiltonian parameters $g_{||} = 2.40–2.44$ and $A_{||} \sim 110$ G observed in Cu-doped TiO_2 has been attributed to the copper linked with H_2O or O_2^-, while the signal with $g_{||} = 2.32$ and $A_| \sim 154$ G has been assigned to Cu^{2+} ion associated with $(OH)^-$ (Altynnikov et al. 2006). The latter parameters are close to the parameters of the s1 component. Therefore, this signal is more likely related to the surface complex containing OH groups. As Figure 2.12 shows, this statement is supported by the correlation of its intensity with the amount of water molecules and OH groups at the surface which were recorded in the infrared attenuated total reflection spectra of the nanocrystals (Korsunska et al. 2017a). Similar dependence is also observed for other components of spectrum I.

Correlation between this EPR spectrum and the presence of surface complexes involving water molecules and/or OH groups is also evidenced by an annealing of the xerogel at 1100°C followed by quenching. This treatment results in the simultaneous appearance of EPR spectrum I and infrared absorption bands related to OH groups and H_2O molecules (Figure 2.13). The same annealing followed by slow cooling does not cause either infrared absorption bands or EPR spectrum I.

Thus, the EPR spectrum I can be attributed to the surface complexes containing copper ions linked with water molecules and/or OH groups.

Note that the decrease of EPR intensity of spectrum I is accompanied by the increase of the absorption band at 600–900 nm related to surface copper ions (either Cu from CuO molecules or Cu_{Zr} substitutions). Since T_c increase above 800°C results in the transformation of this absorption band towards that of crystalline CuO, one can ascribe the absorption band observed for samples calcinated at $T_c = 500–800$°C to dispersed CuO.

As for EPR spectrum II observed at higher (800°C–1000°C) temperatures, it is a set of irregular shape lines in a wide range of magnetic fields. The narrow EPR lines (and therefore the absence of spin-Hamiltonian parameters distribution) indicate that the paramagnetic centres responsible for this spectrum are in regular positions with a stable surrounding.

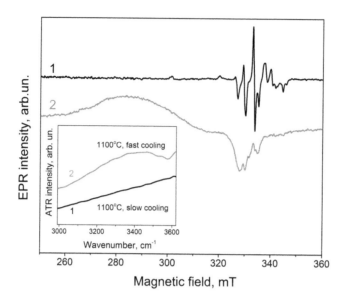

FIGURE 2.13 EPR spectra of Cu-8 sample calcined at 1100°C followed by slow cooling (curve 1) or by quenching (curve 2). The corresponding infrared absorption spectra are shown in the inset.

Besides, the presence of characteristic hyperfine lines allows the reasonable assumption that this spectrum or at least its main part is caused by substitutional Cu^{2+} ions (Cu_{Zr}).

For both types of samples (Cu-1 and Cu-8 samples), signal II shows different behaviours with temperature increase (Figure 2.11), but it has a similar trend with the contribution of monoclinic phase: its intensity initially increases with monoclinic-to-tetragonal (-cubic) ratio and then decreases (Figure 2.14). Since the signal II is observed only when the monoclinic phase appears, it can be assumed to be caused by Cu_{Zr}^{2+} ions in a monoclinic structure.

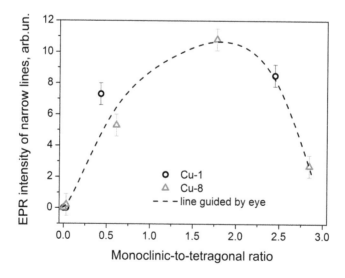

FIGURE 2.14 Variation of the intensity of EPR signal II versus monoclinic-to-tetragonal (-cubic) ratio for Cu-1 (circles) and Cu-8 (triangles) samples.

2.4.5 Copper Redistribution, Phase Transformation and Catalytic Activity of the Powders

The above results provide information on the redistribution of copper between the volume and the surface of the nanocrystals depending on the calcination temperature. The observation of the absorption band caused by oxygen vacancies testifies to Cu presence inside the grains. This is in accordance with the higher intensity of this band in Cu-8 samples as compared with Cu-1 ones. The presence of copper in the bulk of nanocrystals is also confirmed by XRD data, which demonstrate the shift of XRD peak positions to higher diffraction angles as compared to those of Cu-free Y-doped ZrO_2 samples and shift increase with the increase of Cu content.

With T_c rise (in the range of 500°C–700°C) for Cu-8 samples, the absorption band at ~270 nm intensifies, indicating an increase in the number of oxygen vacancies in the nanocrystals. This matches with the enrichment of nanocrystal volume with copper that is also confirmed by an additional shift of XRD peak positions to higher angles and can be assigned to Cu in-diffusion from the Cu-related surface complexes observed in EPR spectra. Indeed, the T_c increase leads to the decrease of corresponding EPR signal intensity due to destruction of these complexes as a result of water or OH group loss. Simultaneously, the intensity of the CuO-related absorption band (in the range of 600–900 nm) increases. This finding can be explained by the following: part of Cu^{2+} ions appeared due to the destruction of surface complexes penetrates additionally into nanocrystals, volume, while another one oxidises, forming CuO molecules.

As T_c increases in the range of 800°C–1000°C, the intensity of the band at ~270 nm reduces. This is consistent with the appearance of the monoclinic phase and its increasing contribution (Figure 2.6). Appearance of the monoclinic ZrO_2 phase can be explained by outward diffusion of Cu, which stimulates outward diffusion of Y from such grains. The Cu out-diffusion is confirmed by the enhancement and high-energy shift of the absorption band related with CuO molecules (in the range $T_c \leq 800°C$) as well as by the appearance of the absorption feature of crystalline CuO (Figure 2.10). At the same time, the decrease of CuO-related absorption band at $T_c > 800°C$ can be caused by yttrium cuprate formation. The decrease of Cu and Y contents in nanocrystals has been observed also by EDX method, proving consequently their segregation at grain boundaries (Figure 2.9). An additional argument for Cu out-diffusion is the shift of XRD peak positions of tetragonal nanocrystals to lower angles (Figure 2.7).

The increase of calcination temperature stimulates the enrichment of grain surface with CuO or pure Cu no matter the cooling rate. However, the higher the cooling rate, the higher the Cu (CuO) content at grain surface can be seen (Korsunska et al. 2017b). This phenomenon is determined by the balance between the processes of "in-diffusion" and "out-diffusion" of copper ions. The calcination temperature increase shifts this balance towards Cu out-diffusion. In this regard, a higher cooling rate (quenching) allows "freezing," the state corresponding to higher T_c and, thus, the higher surface Cu content.

The Cu relocation can also explain the non-monotonic dependence of the EPR spectrum II intensity versus the ratio of monoclinic-to-tetragonal (cubic) phases. In fact, the increase of the signal intensity with this ratio can be caused by the increasing number of nanocrystals with monoclinic structure, while the decreasing of EPR signal II intensity can be assigned to copper out-diffusion from nanocrystals volume (Figure 2.14). The processes of Cu relocation are presented schematically in Figure 2.15.

Thus, the co-precipitation method allows obtaining nanocomposite with copper on the surface and inside of nanocrystals. Variations of calcination temperature and cooling rates can change the copper concentration on the surface and in the volume of nanocrystals, as well as transformation of Cu-related surface entities from complexes containing water molecules or OH group to dispersed and crystalline CuO and Cu clusters.

Cu relocation with the increase of T_c results also in the phase transformation. In fact, the appearance of cubic phase at $T_c = 1000°C–1100°C$ can be caused by the sintering of powder due to the appearance of crystallised CuO at the grain surface. This results in the appearance of tensile stresses in the nanocrystals with tetragonal structure. These stresses are considered to be the reason for cubic phase formation. Besides, the Cu out-diffusion accompanied by Y loss is the reason of t-m transformation (Korsunska et al. 2017a, 2017c).

Because of the absence of a significant amount of cation vacancies and the relatively low temperature of $t(c)$-m phase transformation, the most probable mechanism of Cu out-diffusion is the interstitial one. In this case, Cu atoms shift to interstitials and then move to the surface. It is known that in a number of semiconductor compounds (specifically in II-VI one), Cu can easily shift to interstitial and is then highly mobile (Korsunska et al. 2017c).

The appearance of cation vacancies due to Cu out-diffusion can stimulate Y out-diffusion via cation vacancies that can also be realised at low temperatures. The rough estimation of the activation energy of Cu shift to interstitial sites followed by its diffusion was found to be nearly 1.9 eV (Korsunska et al. 2017c).

The possibility of controlling the Cu-related entities (their amount and nature: complexes containing water molecules, or OH group, CuO molecules or crystalline CuO and Cu clusters) by the calcination temperature and cooling rate can be useful for catalyst preparation.

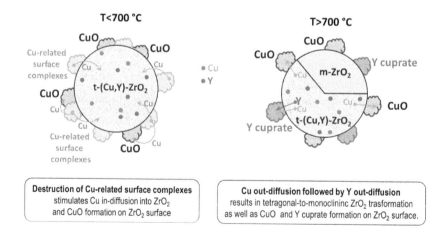

FIGURE 2.15 Schematic presentation of the processes prevailed in Cu-doped YSZ at different temperatures.

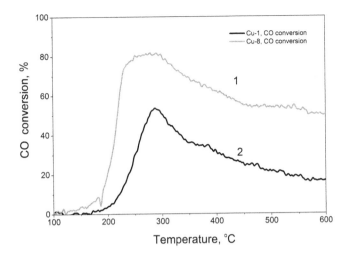

FIGURE 2.16   The dependence of CO conversion on calcination temperature.

The example of the influence of T_c on catalytic activity of composite investigated is shown in Figure 2.16. The reaction of preferential oxidation of CO (so called as PROX-CO) was used as a catalytic test with our Cu-1 samples calcined at 600°C and 800°C. As Figure 2.16 shows, this composite is active to CO oxidation, and CO conversion increases with T_c increase and reaches 80% at 800°C. Such a behaviour correlates with the increase in intensity of the band in diffuse reflectance spectra caused by disperse CuO that is proposed to be responsible for catalytic activity of the composite (Samson et al. 2014; Pakharukova et al. 2015).

2.5 LUMINESCENT PROPERTIES OF ZrO_2-BASED COMPOSITES

The photoluminescence (PL) properties were investigated in ZrO_2 with different additives (Y_2O_3, CuO, Sm_2O_3, Er_2O_3, etc). Nonstoichiometric undoped and Y-doped ZrO_2 demonstrates a broad visible luminescence, which opens the perspective for the use of zirconium oxide in white light-emitting devices (Petrik et al. 1999) as well as in the luminescent oxygen and NO_x sensors (Lakshmi et al. 2011) due to the sensitivity of emission intensity to gas atmosphere. It is known that the photoluminescence (PL) spectra of undoped and yttrium-stabilised ZrO_2 at room temperature contain either one wide unstructured band or several overlapping emitting bands and show the dependence of the PL intensity and peak position on the excitation wavelength (Jia et al. 2006; Nakajima et al. 2006) due to overlapping of different components.

Usually, luminescence is ascribed to native defects: oxygen vacancies or their complexes, located in the volume of powder grains (Jia et al. 2006; Petrik et al. 1999; Korsunska et al. 2014a) or on their surfaces (Nakajima et al. 2006), T-defects in the volume of crystallites (Orera et al. 1990) or unsaturated Zr bonds on their surfaces (surface Zr^{3+} centres) (Jacob et al. 1994), distortions of the lattice with oxygen vacancies (Smits et al. 2011) and complexes of vacancies with impurities. Indeed, the introduction of impurities with a valency less than the valency of Zr^{4+} results in the creation of anionic vacancies. Therefore, it can be expected that such doping will lead to an increase in the intensity of PL bands associated with oxygen vacancies or their complexes with impurities. Indeed, such an effect was observed in Y-doped ZrO_2 (Nakajima et al. 2006).

2.5.1 Rare Earth Co-doped Composites

Besides Y, other rare-earth (RE) elements (Sm, Er, Pr, Eu, Tb, Yb) were used for modification of the luminescence properties of ZrO_2. All of them are usually subvalent and can stabilise the tetragonal or cubic phases (RE can be tetravalent only at a high doping concentration (J.Yuan et al. 1999). Therefore, besides the specific bands corresponding to intra-shell transition, emission caused by oxygen vacancies can also be observed.

The interest in rare-earth ions in zirconia is caused by their emission in the near-infrared range that can be useful for medical application and for communication purposes. Besides, the rare-earth-doped samples may find application in light-emitting devices. On the other hand, ZrO_2 is one of the perspective oxides which can be used as a host material for upconversion luminescence, which can be realised in material co-doped with two RE ions. However, their incorporation results in the formation of oxygen vacancies. In this case, the oxygen vacancies associated mainly with Zr in the vicinity of the RE ions (Veal et al. 1988; Cole et al. 1990) could influence their emission. For example, the energy transfer from the host defect to Sm^{3+} ions was described in De la Rosa et al. (2005).

However, a more impressive effect was observed in Er and Yb co-doped composites (Smits et al. 2014) and consisted of the influence of oxygen vacancies on the energy transfer between ions involved in the upconversion process. In this case, the presence of these vacancies was shown to hamper this process. To reduce the oxygen vacancies' content, additional doping with Nb^{5+} ions was performed in Smits et al. (2014, 2017).

The upconversion process was realised in Er and Yb co-doped zirconia. It was shown (Solis et al. 2010) that the doping with Yb enhances the efficiency of the upconversion process that is caused by the large cross section of Yb optical absorption near ~1 μm and effective energy transfer to Er ions.

The nanopowders of Er- and Yb-doped zirconia (Er_2O_3 (0.01), Yb_2O_3 (0.02)), with the addition of Nb_2O_3 in different content obtained by the sol-gel method, were investigated (Smits et al. 2014). The samples without Nb show mainly the tetragonal phase at different calcination temperatures, while Nb addition results in an appearance of a monoclinic one, wherein the higher the Nb content, the lower the temperature of *t-m* transformation. The reason for this transformation is assigned to the ability of Nb^{5+} (which is the donor) to compensate the charge of trivalent RE ions instead of oxygen vacancies resulting in their content decrease. Besides, the Nb addition was found to strongly influence the efficiency of energy transfer between Yb and Er, resulting in the significant enhancement of the upconversion process (Figure 2.17a,b). The enhancement of the integral intensity of upconversion emission (540–550 nm) was found to be dependent on the Nb content and annealing temperature (Figure 2.17).

In the samples annealed at the temperatures above 1000°C, such emission increased up to twenty times compared to the samples without Nb (Figure 2.18). Because the main upconversion mechanism in the system investigated is energy transfer from Yb^{3+} to Er^{3+}, it can be concluded that Nb doping enhances this process. In spite of the possible stimulation of monoclinic phase formation due to Nb doping, this enhancement cannot be assigned

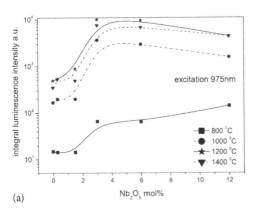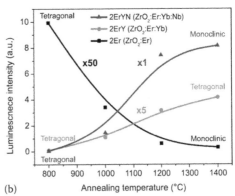

(a)   Nb₂O₅ mol%   (b)   Annealing temperature (°C)

FIGURE 2.17   (a) Integral intensity of the luminescence as a function of Nb content in the Nb-doped ZrO₂:Er, Yb samples. (Reproduced with permission from Smits et al. (2014). Copyright (2014), AIP Publishing LLC.) (b) Upconversion luminescence intensities for samples 2Er, 2ErYb and 2ErYbNb annealed at different temperatures. (Reproduced from *Sci Rep*, Doped zirconia phase and luminescence dependence on the nature of charge compensation. 7, 2017, 44453. 2017. Copyright 2017 Springer Nature under the Creative Commons CC BY License.)

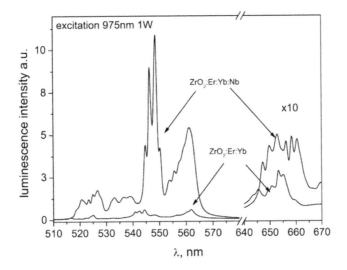

FIGURE 2.18   Upconversion luminescence for Er:Yb-doped zirconia with 3 mol% additional doping by Nb and for Nb-free zirconia. Both samples have been annealed at 1000°C. (Reproduced with permission from Smits et al. (2014). Copyright (2014), AIP Publishing LLC.)

to the simple appearance of this phase, because in the samples without Nb, the *t-m* phase transformation results in the decrease of the upconversion process (Smits et al. 2014). Therefore, it was proposed that the observed effect is caused by the decrease of oxygen vacancies content caused by Nb doping, and the main factor of suppression of upconversion emission is the presence of oxygen vacancies (Smits et al. 2014).

Thus, the obtained results show the influence of the co-doping process on the phase transformation and confirm the previous conclusion that oxygen vacancies suppress the upconversion luminescence.

2.5.2 Y, Cu Co-doped Zirconia-based Composite

Besides the RE ions, the influence of Cu on the luminescence properties of Y and Cu co-doped composite was also investigated. In addition to Cu influence on structural characteristics of composite, one can also expect the appearance of a specific luminescence band caused by copper incorporated in the grains. In general, this band may originate from the intra-shell transition in isolated copper ion or be associated with a complex, which includes, in addition to copper, an oxygen vacancy. In this case, the emission can occur in vacancies, distorted by impurity, or in the impurities themselves when the energy transfer from vacancy to impurity takes place.

Thus, the study of the luminescence of copper-doped composites can confirm Cu localisation inside the grains and clarify the Cu-vacancy interaction. Besides, the copper-related luminescence may be useful for the elaboration of light sources. The luminescence study described below was performed for the same powders which were used for structural and optical investigations described above. The photoluminescence (PL) and PL excitation (PLE), as well as cathodoluminescence (CL) spectra, were measured.

The *photoluminescence* spectra of ZrO_2 doped with yttrium and copper measured at $\lambda_{ex} = 290$ nm are shown in Figure 2.19a. In the samples annealed at 500°C–800°C, they contain two broad bands with the maxima of ~540 nm and ~630 nm, a shoulder in the 470–500 nm region and a long-wave tail similar to that observed in copper-free zirconia. In general, these spectra are similar to those of yttrium-doped samples, which indicates the identity of the emitting centres in these samples.

An increase of annealing temperature to 1000°C leads to a change in the shape of the luminescence spectrum, namely: to the widening of the green band and the formation of a plateau instead of a maximum. This transformation is due, obviously, to the appearance of an additional, shorter-wavelength component of emission. To detect its spectral position, the ratio of spectra for samples annealed at different temperatures was used.

As can be seen from Figure 2.19c, annealing at 900°C does not change the spectrum of luminescence (the ratio of spectra of the samples annealed at 900°C and 800°C is almost

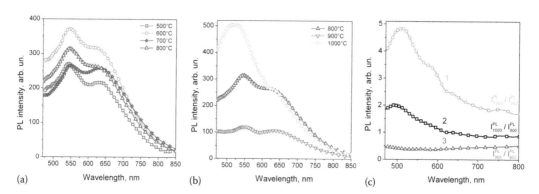

FIGURE 2.19   PL spectra of powders doped with yttrium and copper, annealed at 500°C–800°C (a) and 800°C–1000°C (b) under excitation at $\lambda_{ex} = 290$ nm. The PL spectrum for 800°C in (a) and (b) corresponds to the same sample. Figure (c) shows the ratios of the spectra obtained for the samples annealed at 1000°C and 900°C (1), 1000°C and 800°C (2) and 900°C and 800°C (3).

a straight line). At the same time, the ratio of the spectra of samples annealed at 1000°C and 900°C shows a preferential increase of the two PL bands at ~510 nm and ~580 nm (Figure 2.19c). Comparison with the spectra of samples annealed at 800°C also shows an increase of the short-wave component of the green band and shoulder in the region of 550–600 nm.

The shape of PL spectra of all samples varies with excitation light wavelength. This means that they consist of several overlapping components typically for pure and Y-doped ZrO_2 materials (Petrik et al. 1999; Nakajima et al. 2006). On the other hand, this variation implies that PL bands possess the individual excitation features. Unfortunately, these latter cannot be clearly separated due to significant overlapping of PL bands. However, the variation of excitation wavelength allows determining the spectral range in which the different PL bands can be seen clearly.

This is demonstrated in Figure 2.20a, which presents the PL spectra of the Cu-8 sample annealed at 1000°C. Specifically, when the wavelength of the exciting light (λ_{ex}) decreases, the intensity of the long-wavelength tail of the PL spectra increases, and at $\lambda_{ex} = 230$ nm, the appearance of a band in the red region with a maximum of ~680–700 nm (red band) is observed.

The excitation spectra for different wavelengths of emission (λ_{PL}), as well as the absorption spectrum for copper-doped samples, are shown in Figure 2.20b. As can be seen, for all wavelengths of registration in the region 450–570 nm, in addition to excitation at the edge of band-to-band absorption, there is an impurity maximum at ~280–290 nm, which is localised in the region of light absorption by oxygen vacancies (Figure 2.20b, curve 7).

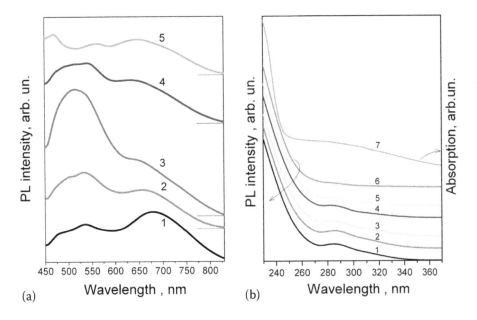

FIGURE 2.20 (a) PL spectra of the sample doped with yttrium and copper and annealed at 1000°C, measured under excitation at $\lambda_{ex} = 230$ (1), 270 (2), 290 (3), 330 (4) and 370 (5) nm. The spectra are shifted vertically for clarity; (b) PLE spectra of the same sample for different registration wavelength: $\lambda_{PL} = 480$ (1), 510 (2), 530 (3), 570 (4), 630 (5) and 700 (6) nm. Curve 7 corresponds to absorption spectrum. All PLE spectra are normalised and shifted vertically.

However, its intensity is the highest for emission at 510 nm, i.e. for an additional green band. Unfortunately, the other possible individual maxima for different PL bands were not clearly observed, obviously at their lower intensity than peak, at ~280–290 nm. At the same time, for the red emission, this impurity maximum is absent, and light excitation is dominated by the ZrO_2 band-to-band absorption.

The intensive band-to-band excitation of luminescence can be realised under an electron beam irradiation. With such an excitation, the *cathodoluminescence* (CL) spectra of Cu-doped Y-stabilised samples, likewise their PL spectra, depend on the annealing temperature. The CL spectra of samples annealed at 500°C–900°C are similar to the CL spectra of the powders doped with yttrium (Figure 2.21a) and contained at 300 K two broad bands with maxima at ~430 and ~700 nm (curve 1). The peak position of the latter is close to the peak position of the red band in the PL spectra. However, in the CL spectra, its intensity significantly exceeds the intensity of the blue band. The blue band has a long-wavelength tail, indicating the presence of a number of other bands in the blue-green region, which are often observed in photo-, X-rays and cathodoluminescence spectra (Petrik et al. 1999; Nakajima et al. 2006; Smits et al. 2011; Grigorjeva et al. 2009).

Cooling down to 77 K increases the intensity of both bands but mainly the red one (Figure 2.21a, curve 2). Thus, the emission shows a thermal quenching in the range of 77–300 K, the peak position and the shape of the red band being unchanged.

At the same time, the sample annealed at 1000°C shows at 300 K the significant increase of green bands contribution, mainly the band at ~510 nm (Figure 2.21b, curve 1). The dependencies of the intensity for all emission bands on the value of electron beam current are sublinear, indicating that the number of excited centres is close to their concentration.

At 77 K in the samples annealed at 1000°C, the intensity of all the bands increases, mainly of the red one (Figure 2.21b, curve 2). At the same time, the broad band in the blue-green region, which is observed at room temperature, is divided into a blue one (with a maximum of 430 nm) and a green one (with a maximum of 520 nm). The shift of the peak

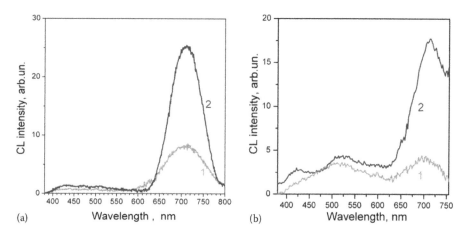

FIGURE 2.21   CL spectra of the samples annealed at $T_c < 900°C$ (a) and at $T_c = 1000°C$ (b), measured at 300 (1) and 77 K (2). Excitation of luminescence was carried out at electron acceleration voltage $E = 75$ kV and current $I = 0.12$ mA.

position of the latter from 510 nm to 520 nm is caused apparently by the increase of the contribution of the shoulder at 550–600 nm.

As was mentioned above, radiative transitions in the range of 380–600 nm in undoped and yttrium-doped ZrO_2 are attributed to their own defects (oxygen vacancies – F and F^+ centres, etc.), located in the volume of grains or on their surface. The presence of an impurity maximum in the region of light absorption by oxygen vacancies in their PLE spectra proves the relation of the corresponding emission centres with vacancies or with impurity-vacancies complexes. Their excitation can occur both due to capturing free carriers (excitation by an electron beam and short-wave photoexcitation) and as a result of light absorption by vacancies. Emission can be a consequence of intra-shell transitions in vacancies or the carrier recombination in impurities (in the case of a complex of impurity with vacancy). In the latter case, the emission can be excited due to the capture of free carriers or as a result of energy transfer from oxygen vacancies.

It should be noted that the red band, which doesn't have an impurity PLE maximum, was observed both in Y-doped and Y, Cu co-doped samples. As is shown in the analysis of CL spectra of Y-doped samples (Korsunska et al. 2014b, 2015a,b), the corresponding emitting centre has at least two energy levels in the forbidden band, and the radiative transition is considered to be intra-defect. This is in agreement with the independence of this band's peak position of temperature in Korsunska et al. (2014b). Because in the excitation spectra of this band there is no impurity maximum related to oxygen vacancies (Figure 2.20), it can proposed that this band is due to the intra-defect transition in the impurities, the nature of which still needs to be clarified and which do not lead to the formation of oxygen vacancies (for example, the impurity of IV group).

To ascertain their origin, the information on the crystalline structure of the investigated powders described above can be used. The samples annealed at temperatures below 900°C have a predominantly tetragonal structure, while at higher temperatures a monoclinic phase is observed, the contribution of which increases with T_c. Therefore, we can assume that the appearance of additional bands in PL and CL spectra of the samples doped with copper is caused by copper-related emitting centres in the nanocrystals of monoclinic phase. In these nanocrystals, the copper concentration is lower than in the nanocrystals of the tetragonal phase, and it is rather the impurity than component of solid solution.

As was mentioned above, in the EPR spectra of copper-doped samples annealed at temperatures above 800°C, a series of narrow lines in a wide region of magnetic fields is observed. They were assigned to the substitutional copper ions Cu^{2+} (Cu_{Zr}) in nanocrystals of the monoclinic phase (Korsunska et al. 2017b) and can be considered as the origin of emission at 510 nm and 550–600 nm. This assumption is confirmed by the following experiment.

The sample with an intensive band at 510 nm was additionally annealed at 1100°C for 20 minutes and quickly cooled to room temperature. Before the additional annealing, EPR spectra showed a signal from the Cu_{Zr} centres (Figure 2.22a, curve 1). After the annealing, the intensity of this signal decreased, and another broad signal caused by copper surface complexes containing OH groups and water molecules appeared (Korsunska et al. 2017b) (Figure 2.22a, curve 2). This transformation of the EPR spectra can be explained by the

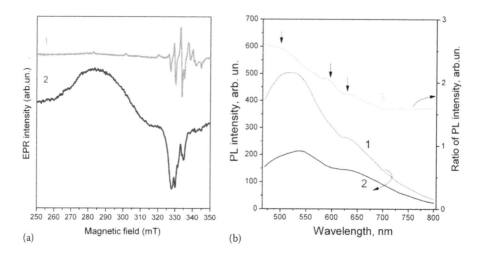

FIGURE 2.22   EPR (a) and PL (b) spectra of the samples doped with yttrium and copper before (1) and after annealing at 1100°C followed by rapid cooling (2). Curve 3 in Fig. b) is the ratio of spectra 1 and 2.

diffusion of copper atoms from the volume of nanocrystals to their surface. Simultaneously with the transformation of the EPR spectra, a decrease in the intensity of the PL bands at 510 nm and 580 nm was observed. This can be seen from the ratio of the spectra before and after additional annealing (Figure 2.22b, curves 1, 2). It should be noted that in addition to these two bands, a decrease in the intensity of the band at 630 nm, which is often ascribed to oxygen vacancies, also occurred. It is more clearly seen from the ratio of spectra 1 and 2 from Figure 2.22b, shown as curve 3. The decrease from 600 nm band intensity can be understood based on the fact that a decrease in the concentration of Cu_{Zn} centres can lead to a decrease in the concentration of compensating their oxygen vacancies.

The presence of two emission bands at 510 nm and 580 nm can be attributed to intra-shell transitions in copper ions. The appearance of a Cu-related band in the green region of the spectrum was also observed in Cu-doped ZnO (Alivov et al. 2004).

It should be noted that the number of bands and their spectral position in ZrO_2 are similar to the emission in copper vapour lasers. If we take into account that the emission of copper-doped powders is probably due to Cu_{Zn}^{2+} ions, the realisation of transitions similar to the radiative transitions in the lasers can occur as follows. Under equilibrium conditions, one electron in d-shell of Cu_{Zn}^{2+} ion is absent, which is equivalent to the excited state of the copper atom in the lasers, but, unlike this, there are no electrons in the next shell. However, under excitation by the light (in particular from the range of intrinsic absorption), a nonequilibrium electron can be captured on this shell, and its subsequent transition to a level of the d-shell will result in emission. The initial state is restored after the capture of a nonequilibrium hole appeared under excitation, which means the completion of the recombination process.

As PLE spectra show, Cu-related PL bands can be excited not only by free carriers but also through oxygen vacancies that testify to the location of Cu_{Zr} near these vacancies. The proximity of oxygen vacancies to copper ions facilitates the excitation transfer from vacancy to emitting centre. It should be noted that the impurity peak in PLE spectra is narrower than the oxygen-related absorption band, stimulated by impurity incorporation in the grains. Thus, the investigation of PLE spectra can provide information on the vacancies which are nearest to impurity. Other vacancies can be more distant and can be useful for oxygen ion conductivity.

2.6 CONCLUSION

As the presented results show, the addition of several impurities allows essential modification of the properties of the ZrO_2-based composite as compared to pure or one-additive-contained material. Specifically, structural, surface, optical and luminescence properties, as well as ionic or electron conductivity, can be improved. The choice of dopants, variation of their concentration and annealing temperature assert precise enough control of the content of oxygen vacancies that can be used for the enhancement of ionic conductivity (if vacancies content increases) or upconversion process (if their content decreases), the type and content of surface entities that can be useful for control of catalytic activity and antibacterial properties, etc. All these properties can be used for different applications: in solid oxide fuel cells, oxygen sensors, catalysis, for communication purposes, medical application, in light-emitting devices, etc.

REFERENCES

Abbas, H. A., Argirusis, C. and Kilo, M. 2011. Preparation and conductivity of ternary scandia-stabilised zirconia. *Solid State Ionics* 184:6–9.

Alivov, Y. I., Chukichev, M. K. and Nikitenko, V. A. 2004. Green luminescence band of zinc oxide films copper-doped by thermal diffusion. *Semiconductors* 38:31–5.

Altynnikov, A. A., Tsikoza, L. T. and Anufrienko, V. F. 2006. Ordering of Cu(II) ions in supported copper-titanium oxide catalysts. *J Struct Chem* 47:1161–9.

Arachi, Y., Asai, T., Yamamoto, O. and Tamakoshi, C. 2001. Electrical conductivity of ZrO_2-Sc_2O_3 doped with HfO_2, CeO_2 and Ga_2O_3. *J Electrochem Soc* 148:A520–A523.

Asadi, S., Abdizadeh, H. and Vahidshad, Y. 2012. Effect of crystalline size on the structure of copper doped zirconia nanoparticles synthesized via sol-gel. *J Nanostructures* 2:205–212.

Asadikiya, M. and Zhong, Y. 2018. Oxygen ion mobility and conductivity prediction in cubic yttria-stabilized zirconia single crystals. *J Mater Sci* 56:1699–1709.

Badwal, S. P. S. 1992. Zirconia-based solid electrolytes: Microstructure, stability and ionic conductivity. *Solid State Ionics* 52:23–32.

Bhagwat, M., Ramaswamy, A. V., Tyagi, A. K. and Ramaswamy, V. 2003. Rietveld refinement study of nanocrystalline copper doped zirconia. *Mater Res Bull* 38:1713–24.

Bouwmeester, H. J. M. and Burggraaf, A. J. 1996. Dense ceramic membranes for oxygen separation. In *Fund Inorg Memb Sci Tech*, ed. A. Burggraaf, and L. Cot, 435–528. Elsevier: Amsterdam.

Chibaa, R., Yoshimuraa, F., Yamakia, J., Ishiib, T., Yonezawac, T. and Endouc, K. 1997. Ionic conductivity and morphology in Sc_2O_3 and Al_2O_3 doped ZrO_2 films prepared by the sol-gel method. *Solid State Ionics* 104:259–266.

Cole, M., Catlows, C. R. and Dragu, J. P. 1990. EXAFS studies of doped-ZrO_2 systems. *J Phys Chem Sol* 51:507

Etsell, T. H. and Flengas S. N. 1970. Electrical properties of solid oxide electrolytes. *Chem Rev* 70:339–376.

Fabris, S., Paxton, A. T. and Finnis, M. W. 2002. A stabilisation mechanism of zirconia based on oxygen vacancies only. *Acta Mater* 50:5171–8.

Fidelus, J. D., Lojkowski, W., Millers, D., Smits, K. and Grigorjeva, L. 2009. Advanced nanocrystalline ZrO_2 for optical oxygen sensors. *IEEE Sensors* 9:1268–1272.

Fischer, S., Fleischer, M., Magori, E., Pohle, R. and Straub, N. 2017. Gas sensor and method for detecting oxygen. US Patent, Pub. No. 2017/0227487 A1, Pub. Date: 10.08.2017. Washington, DC: U.S. Patent and Trademark Office.

Garvie, R. C. and Goss, M. F. 1986. Intrinsic size dependence phase transformation temperature in zirconia microcrystals. *J Mater Sci* 21:1253–7.

Goff, J. P., Hayes, W., Hull, S., Hutchings, M. T. and Clausen, K. N. 1999. Defect structure of yttria-stabilized zirconia and its influence on the ionic conductivity at elevated temperatures. *Phys Rev B* 59:14202–19.

Gorban, O., Danilenko, I., Gorban, S., Volkova, G., Glazunova, V. and Konstantinova, T. 2017. $Ag(Ag_2O)$-ZrO_2-Y_2O_3 photosensitive composites: Influence of synthesized routes on structure and properties. *Photochem Photobiol Sci* 16:53–59.

Grigorjeva, L., Millers, D., Kalinko, A., Pankratov, V. and Smits, K. 2009. Time-resolved cathodoluminescence and photoluminescence of nanoscale oxides. *J Eur Ceram Soc* 29:255–9.

Gross, M. D., Vohs, J. M. and Gorte R. J. 2007. A study of thermal stability and methane tolerance of Cu-based SOFC anodes with electrodeposited Co. *Electrochim Acta* 52:1951–7.

Habibi, M. H., Wang, L., Liang J. and Guo, S. M. 2013. An investigation on hot corrosion behavior of YSZ-Ta_2O_5 in Na_2SO_4+V_2O_5 salt at 1100°C. *Corros Sci* 75:409–14.

Hearing, C., Roosen, A., Schichil, H. and Schnoller, M. 2005. Degradation of the electrical conductivity in stabilized zirconia system Part II: Scandia-stabilized zirconia. *Solid State Ionics* 176:261–8.

Hong, L., Hu, J.-M., Gerdes, K. and Chen, L.-Q. 2015. Oxygen vacancy diffusion across cathode/electrolyte interface in solid oxide fuel cells: An electrochemical phase-field model. *J Power Sources* 287:396–400.

Inaba, H. and Tagawa, H. 1996. Ceria-based solid electrolytes. *Solid State Ionics* 83:1–16.

Izu, N., Nishizaki, S., Shin, W., Itoh, T., Nishibori, M. and Matsubara, I. 2009. Resistive oxygen sensor using ceria-zirconia sensor material and ceria-yttria temperature compensating material for lean-burn engine. *Sensors (Basel, Switzerland)* 9:8884–95.

Jacob, K. H., Knozinger, E. and Benfer, S. 1994. Chemisorption of H_2 and H_2-O_2 on polymorphic zirconia. *J Chem Soc Faraday Trans* 90:2969–75.

Jayakumar, S., Ananthapadmanabhan, P. V., Thiyagarajan, T.K., Perumal, K., Mishra, S. C., Suresh, G., Su, L. T. and Tok, A. I. Y. 2013. Nanosize stabilisation of cubic and tetragonal phases in reactive plasma synthesized zirconia powders. *Mat Chem Phys* 140:176–82.

Jia, R., Yang, W., Bai, Y. and Li, T. 2006. Upconversion photoluminescence of ZrO_2:Er^{3+} nanocrystals synthesized by using butadinol as high boiling point solvent. *Opt Mater* 28:246–9.

Kajiyama, N. and Nakamura, T. 2015. Electrode for use in gas sensor and gas sensor element using the same. US Patent, Pub. No. 2015/0293051 A1, Pub. Date: 15.10.2015. Washington, DC: U.S. Patent and Trademark Office.

Kharton, V. V., Marques, F. M. B. and Atkinson, A. 2004. Transport properties of solid oxide electrolyte ceramics: A brief review. *Solid State Ionics*, 174:135–49.

Kharton, V. V., Naumovich, E. N. and Vecher A. A. 1999. Research on the electrochemistry of oxygen ion conductors in the former Soviet Union. I. ZrO_2-based ceramic materials. *J. Solid State Electrochem* 3:61–81.

Killner, J. A. and Brook, R. J. 1982. A study of oxygen ion conductivity in doped nonstoichiometric oxides. *Solid State Ionics* 6:237–52.

Kirm, M., Aarik, J. and Sildos, I. 2005. Thin films of HfO_2 and ZrO_2 as potential scintillators. *Nucl Instrum Methods Phys Res A* 537:251–5.

Kilo, M., Argirusis, C., Borchardt, G. and Jackson, R. A. 2003. Oxygen diffusion in yttria stabilised zirconia – experimental results and molecular dynamics calculations. *Phys Chem Chem Phys* 5:2219–24.

Kong, L., Bi, Q., Zhu, S., Qiao, Z., Yang, J. and Liu, W. 2014. Effect of CuO on self-lubricating properties of $ZrO_2(Y_2O_3)$-Mo composites at high temperatures. *J Europ Ceram Soc* 34:1289–96.

Korsunska, N., Baran, M., Zhuk, A., Polishchuk, Y., Stara, T., Kladko, V., Bacherikov, Y., Venger, Y., Konstantinova, T. and Khomenkova, L. 2014a. Role of paramagnetic defects in light emission processes in Y-doped ZrO_2 nanopowder. *Mater Res Express* 1:045011.

Korsunska., N., Papusha, V., Kolomys, O., Strelchuk, V., Kuchuk, A., Kladko, V., Bacherikov, Y., Konstantinova, T. and Khomenkova, L. 2014b. Nanostructured Y-doped ZrO_2 powder: Peculiarities of light emission under electron beam excitation. *Phys Stat Sol C* 11:1417–22.

Korsunska, N., Stara, T., Khomenkova, L., Polishchuk, Y., Kladko, V., Michailovska, K., Kharchenko, M. and Gorban, O. 2015a. Effect of Cu- and Y-Codoping on structural and luminescent properties of zirconia based nanopowders. *ECS Trans.* 66:313–9.

Korsunska, N., Baran, M., Polishchuk, Y., Kolomys, O., Stara, T., Kharchenko, M., Gorban, O., Strelchuk, V., Venger, Y., Kladko, V. and Khomenkova, L. 2015b. Structural and luminescent properties of (Y,Cu)-codoped zirconia nanopowders. *ECS Journal of Solid State Science and Technology* 4:N103–N110.

Korsunska, N., Baran, M., Vorona, I., Nosenko, V., Lavoryk, S., Portier, X. and Khomenkova, L. 2017a. Impurity-governed modification of optical and structural properties of ZrO_2-based composites doped with Cu and Y. *Nanoscale Res Lett* 12:157.

Korsunska, N., Baran, M., Vorona, I., Nosenko, V., Lavoryk, S., Polishchuk, Y., Kladko, V., Portier, X. and Khomenkova, L. 2017b. Effect of cooling rate on dopant spatial localization and phase transformation in Cu-doped Y-stabilized ZrO_2 nanopowders. *Phys Stat Sol S* 14:1700183.

Korsunska, N., Polishchuk, Y., Kladko, V., Portier, X. and Khomenkova, L. 2017c. Thermo-stimulated evolution of crystalline structure and dopant distribution in Cu-doped Y-stabilized ZrO_2 nanopowders. *Mater Res Express* 4:035024.

Korsunska N., Baran, M., Papusha, V., Lavoryk, S., Marchylo, O., Michailovska, K., Melnichuk, L., Melnichuk, A. and Khomenkova, L. 2018. The peculiarities of light absorption and light emission in Cu-doped Y-stabilized ZrO_2 nanopowders, *App Nanosci* submitted.

Kumar, A., Jaiswal, A., Sanbui, M. and Omar, S. 2017. Oxygen-ion conduction in scandia-stabilized zirconia-ceria solid electrolyte (xSc_2O_3-$1CeO_2$-$(99-x)ZrO_2$, $5 \leq x \leq 11$). *J Am Ceram Soc* 100:659–68.

Lakshmi, J. S., John Berlin, I., Daniel G. P., Thomas P. V. and Joy, K. 2011. Effect of calcination atmosphere on photoluminescence properties of nanocrystalline ZrO_2 thin films prepared by sol-gel dip coating method. *Physica B* 406:3050–5.

Lee, D.-S., Kim, W. S., Choi. S. H., Kim, J., Lee, H.-W. and Lee, J.-H. 2005. Characterization of ZrO_2 co-doped with Sc_2O_3 and CeO_2 electrolyte for the application of intermediate temperature SOFCs. *Solid State Ionics* 176:33–9.

Liu, W., Ou, G., Yao, L., Nishijima, H. and Pan, W. 2017. Enhanced ionic conductivity in phase stabilized yttria-doped zirconia nanowires. *Solid State Ionics* 308:34–9.

Luo, J., Almond, D. P. and Stevens, R. 2000. Ionic mobilities and association energies from an analysis electrical impedance of ZrO_2-Y_2O_3 alloys. *J Am Ceram Soc* 83:1703–8.

Ma, Z.-Y., Yang, C., Wie, W., Li, W.-H. and Sun, Y.-H. 2005. Catalytic performance of copper supported on zirconia polymorphs for CO hydrogenation. *J Mol Catal A: Chem* 231:75–81.

Mahato, N., Banerjee, A., Gupta, A., Omar, S. and Balani, K. 2015. Progress in material selection for solid oxide fuel cell technology: A review. *Prog Mater Sci* 72:141–337.

Massalski, T. B., Okamoto, H., Subramanian, P. R. and Kacprzak, L. 1990. *Binary Alloy Phase Diagrams*. 2nd Edition. ASM International, Materials Park, OH. ISBN: 978-0-87170-403-0

Minh, N. Q. 1993. Ceramic fuel cells. *J Am Ceram Soc* 76:563–588.

Mogensen, M., Sammes, N. M. and Tompsett, G. A. 2000. Physical, chemical and electrochemical properties of pure and doped ceria. *Solid State Ionics* 129:63–94.

Nakajima, H. and Mori, T. 2006. Photoluminescence excitation bands corresponding to defect states due to oxygen vacancies in yttria-stabilized zirconia. *J Alloys Compounds* 408–412:728–31.

Orera, V. M., Merino, R. I., Chen, Y., Cases, R. and Alonso, P. J. 1990. Intrinsic electron and hole defects in stabilized zirconia single crystals. *Phys Rev B* 42:9782–9.

Omar S., Belda, A., Escardino, A. and Bonanos, N. 2011. Ionic conductivity ageing investigation of 1Ce10ScSZ in different partial pressures of oxygen. *Solid State Ionics* 184:2–5.

Padture, N. P., Gell, M. and Jordan, E. H. 2002. Thermal barrier coatings for gas-turbine engine applications. *Science* 296:280–4.

Pakharukova, V. P., Moroz, E. M., Kriventsov, V. V., Larina, T. V., Boronin, A. I., Dolgikh, L. Y. and Strizhak, P. E. 2009. Structure and state of copper oxide species supported on yttria-stabilized Zirconia. *J Phys Chem C* 113:21368–75.

Pakharukova, V. P., Moroz, E. M., Zyuzin, D. A., Ishchenko, A. V., Dolgikh, L. Y. and Strizhak, P. E. 2015. Structure of copper oxide species supported on monoclinic zirconia. *J Phys Chem C* 119:28828–35.

Pugachevskii, M. A., Zavodinskii, V. G. and Kuz'menko, A. P. 2011. Dispersion of zirconium dioxide by pulsed laser radiation. *Tech Phys* 56:254–8.

Radhakrishnan, J. K., Kamble, S. S., Krishnapur, P. P., Padaki, V. C. and Gnanasekaran, T. 2012. Zirconia oxygen sensor for aerospace applications. *Proceedings of IEEE Sixth International Conference on Sensing Technology (ICST)*, 18–21 December 2012, Kolkata, India, p. 714–7.

Petrik, N. G., Tailor, D. P. and Orlando, T. M. 1999. Laser-stimulated luminescence of yttria-stabilized cubic zirconia crystals. *J Appl Phys* 85:6770–6.

Ran, S., Winnubst, L., Blank, D. H. A., Pasaribu, H. R., Sloetjes, J.-W. and Schipper, D. J. 2007. Effect of microstructure on the tribological and mechanical properties of CuO-doped 3Y-TZP ceramics. *J Am Ceram Soc* 90:2747–52.

Reiter, W. and Seyr, P. 2016. Gas sensor element. US Patent, Pub. No. 2016/0161441 A1, Pub. Date: 09.06.2016. Washington, DC: U.S. Patent and Trademark Office.

Restivo, T. A. G. and Mello-Castanho, S. R. H. 2009. Sintering studies on Ni-Cu-YSZ SOFC anode cermet processed by mechanical alloying. *J Therm Anal Calorim* 97:775–80.

Riley, B. 1990. Solid oxide fuel cells – the next stage. *J Power Sources* 29:223–237.

De la Rosa, E., Salas, P., Desirena, H., Angeles, C. and Rodríguez, R. A. 2005. Strong green upconversion emission in ZrO_2:Yb^{3+}–Ho^{3+} nanocrystals. *Appl Phys Lett* 87:241912.

Sammes, N. M., Tompsett G. A., Näfe, H. and Aldinger, F. 1999. Bismuth based oxide electrolytes – structure and ionic conductivity. *J Europ Ceram Soc* 19:1801–26.

Samson, K., Śliwa, M., Socha, R. P., Góra-Marek, K., Mucha, D., Rutkowska-Zbik, D., Paul, J.-F., Ruggiero-Mikołajczyk, M., Grabowski, R. and Słoczyński, J. 2014. Influence of ZrO_2 structure and copper electronic state on activity of Cu/ZrO_2 catalysts in methanol synthesis from CO_2. *ACS Catal* 4:3730–41.

Schulz, U., Fritscher, K. and Peters, M. 1996. EB-PVD Y_2O_3- and CeO_2-Y_2O_3-stabilized zirconia thermal barrier coatings – crystal habit and phase composition. *Surf Coat Technol* 82:259–69.

Singhania, A. and Gupta, S. M. 2017. Low-temperature CO oxidation over Cu/Pt co-doped ZrO_2 nanoparticles synthesized by solution combustion. *Beilstein J Nanotechnol* 8:1546–52.

Sherif, A.S. and Hachtmann, J. E. 1980. Stable copper zirconium complex salt solutions for enhancing the resistance to rot of cotton fabrics. US Patent No 4,200,672. Washington, DC: U.S. Patent and Trademark Office.

Smits, K., Grigorjeva, L., Millers, D., Sarakovskis, A., Grabis, J. and Lojkowski, W. 2011. Intrinsic defect related luminescence in ZrO_2. *J Lumin* 131:2058–62.

Smits, K., Sarakovskis, A., Grigorjeva, L., Millers, D. and Grabis, J. 2014. The role of Nb in intensity increase of Er ion upconversion luminescence in zirconia. *J Appl Phys* 115:213520.

Smits, K., Olsteins, D., Zolotarjovs, A., Laganovska, K., Millers, D., Ignatans, R. and Grabis, J. 2017. Doped zirconia phase and luminescence dependence on the nature of charge compensation. *Sci Rep* 7:44453.

Solis, D., De la Rosa, E., Meza, O., Diaz-Torres, L. A., Salas, P. and Angeles-Chaves, C. 2010. Role of Yb^{3+} and Er^{3+} concentration on the tunability of green-yellow-red upconversion emission of codoped ZrO_2:Yb^{3+}–Er^{3+} nanocrystals. *J Appl Phys* 108:023103.

Soo, Y. L., Chen, P. J., Huang, S. H., Shiu, T. J., Tsai, T. Y., Chow, Y. H., Lin, Y. C., Weng, S. C., Chang, S. L., Wang, G., Cheung, C. L., Sabirianov, R. F., Mei, W. N., Namavar, F., Haider, H., Garvin, K. L., Lee, J. F. and Chu, P. P. 2008. Local structures surrounding Zr in nanostructurally stabilized cubic zirconia: Structural origin of phase stability. *J Appl Phys* 104:113535.

Sun, Y. and Sermon, P. A. 1994. Evidence of a metal-support in sol-gel derived Cu-ZrO_2 catalysts for CO hydrogenation. *Catalysis Lett* 29:361–369.

Sun, C. and Stimming, U. J. 2007. Recent anode advances in solid oxide fuel cells. *J Power Sources* 171:247–260.

Tok, A. I. Y., Boey, F. Y. C., Du, S. W. and Wong, B. K. 2006. Flame spray synthesis of ZrO_2 nanoparticles using liquid precursors. *Mat Sci Eng B* 130:114–119.

Vasylyev, O. D., Brodnikovskyi, Y. M., Brychevskyi, M. M., Polishko, I. O., Ivanchenko, S. E. and Vereshchak, V. G. 2018. From powder to power: Ukrainian way. *SF J Material Chem Eng* 1:1001.

Veal B. W., McKale A. G., Paulikas A. P., Rothman S. J. and Nowicki L. J. 1988. EXAFS study of yttria stabilized cubic zirconia. *Physica B+C*, 150:234–40.

Villella, P., Conradson, S. D., Espinosa-Faller, F. J., Foltyn, S. R., Sickafus, K. E., Valdez, J. A. and Degueldre, C. A. 2001. Local atomic structure in cubic stabilized zirconia. *Phys Rev B* 64:101104.

Wang, L.-C., Liu, Q., Chen, M., Liu, Y.-M., Cao, Y. and Fan, K.-N. 2007. Structural evolution and catalytic properties of nanostructured Cu/ZrO_2 catalysts prepared by oxalate gel-coprecipitation technique. *J Phys Chem C* 111:16549–57.

Wang, F., Banerjee, D., Liu, Y., Chen, X. and Liu, X. 2010. Upconversion nanoparticles in biological labeling, imaging, and therapy. *The Analyst* 135:1839–54.

Winnubst, L., Ran, S., Speets, E. A. and Blank, D. H. A. 2009. Analysis of reactions during sintering of CuO-doped 3Y-TZP nano-powder composites. *J Europ Ceram Soc* 29:2549–57.

Yamamoto, O., Arachi, Y., Sakai, H., Takeda, Y., Imanishi, N., Mizutani, Y., Kawai, M. and Nakamura, Y. 1998. Zirconia based oxide ion conductors for solid oxide fuel cells. *Ionics* 4:403–8.

You, R., Jing, G., Yu, H. and Cui, T. 2017. Flexible mixed-potential-type (MPT) NO_2 sensor based on an ultra-thin ceramic film. *Sensors (Basel, Switzerland)* 17:1740.

Yuan, J., Hirayama, T., Ikuhara, Y. and Sakuma, T. 1999. Electron energy loss spectroscopy study of cerium stabilised zirconia: An application of valence determination in rare earth systems. *Micron* 30:141–5.

Zhang, C., Li, C.-J., Zhang, G., Ning, X.-J., Li, C.-X., Liao, H. and Coddet, C. 2007. Ionic conductivity and its temperature dependence of atmospheric plasma-sprayed yttria stabilized zirconia electrolyte. *Mater Sci Eng B* 137:24–30.

Zhang, Y., Hu, L., Li, H.K. and Chen, J. 2008a. Densification and phase transformation during pressureless sintering of nanocrystalline ZrO_2-Y_2O_3-CuO ternary system. *J Am Ceram Soc* 91:1332–4.

Zhang, Y. S., Hu, L. T., Zhang, H., Chen, J. M. and Liu, W. M. 2008b. Microstructural characterization and crystallization of ZrO_2-Y_2O_3-CuO solid solution powders. *J Mater Proc Technol* 198:191–4.

Zhang, Y., Chen, C., Lin, X., Li, D., Chen, X., Zhan, Y. and Zheng, Q. 2014. CuO/ZrO_2 catalysts for water-gas shift reaction: Nature of catalytically active copper species. *Int J Hydrogen Energy* 39:3746–54.

CHAPTER 3

Semiconductor Nanocrystals Embedded in Polymer Matrix

Lyudmyla Borkovska

CONTENTS

3.1 BASIC PROPERTIES OF THE QD-POLYMER COMPOSITES AND THE MAIN FIELDS OF THEIR APPLICATION

Over the last decades, the nanocomposite system of polymer embedded with inorganic semiconductor nanoparticles has attracted a lot of attention due to high potential in the development of a variety of novel photonic and photovoltaic (PV) nanostructures and devices. The nanocomposites combine the advantages of both material classes. In particular, size-dependent bandgap, high molar extinction coefficient, solution processability, large dipole moments, photo-stability and high optical absorption coefficients of semiconductor nanoparticles are complemented with the high technological capabilities of polymers, including inherent compositional flexibility, ease of processing, low cost, the possibility of manipulation of their properties through chemical modifications, etc.

Semiconductor nanocrystals whose radii are smaller than the bulk exciton Bohr radius, or *quantum dots* (QDs), constitute a new class of materials intermediate between molecular and bulk forms of matter. The density of states of semiconductor QDs is similar to the atomic density of states. The atomic-like behaviour leads to a relatively easy generation of individual excitons, individual spins of quantum dots and great coherence times of these excitations (Nirmal et al. 1994, 1995; Kroutvar et al. 2004; Flisskowski et al. 2001;

Crooker et al. 2003; Gupta et al. 2002). Quantum confinement of both electrons and holes in all three dimensions leads to an increase of the effective band gap of the material with the decrease of crystallite size. As a result, both the optical absorption and luminescence spectra of QDs shift to higher energies as the size of the nanocrystals decreases.

Until now, much progress has been made in the synthesis and characterisation of monodisperse nanocrystals of a wide variety of semiconductors, such as II-VI, III-V, I-III-VI, Si, carbon, etc. (Gaponic et al. 2010; Harris and Bawendi 2012; Konstantanos et al. 2013; Lesnyak et al. 2013; Grim et al. 2015; Shen et al. 2016). The most widely used and studied QDs consist of CdSe or CdTe nanocrystals. These QDs exhibit intense size-dependent excitonic emissions that span the entire visible spectrum. In particular, by changing a diameter of CdTe nanocrystals from about 2.0 nm to 5.5 nm, the emission peak can be shifted from green to deep red (Rogach et al. 2007) (Figure 3.1). The II-VI nanocrystals are characterised by a narrow emission bandwidth of 30–45 nm and broad absorption spectra, so that several QDs with different emission wavelengths can be excited at a single wavelength and can be spectrally well resolved. However, due to the high surface-to-volume ratio, the surface properties have a significant effect on the QD emission (Dabbousi et al. 1997; Boles et al. 2016). To improve the luminescence quantum efficiency, a growing of a thin inorganic shell of the wide-band gap semiconductor around the particles had been proposed (Dabbousi et al. 1997; Kuno et al. 1997; Peng et al. 1997; Hines et al. 1996). Covering the CdSe and CdTe QDs with ZnS nanoshells, one can fabricate the so-called core-shell CdSe/ZnS and CdTe/ZnS QDs (Peng et al. 1997; Talapin et al. 2004; Cingarapu et al. 2012). The core-shell CdTe/CdS QDs (Gu et al. 2008) with photoluminescence (PL) efficiency approaching 75% have been reported. For alloyed graded CdSe-ZnS/ZnS core/shell QDs, the absolute PL quantum yield (QY) can reach 88% when an optimal thickness is chosen for the outer shell (Fu et al. 2017).

Embedding colloidal QDs in a polymer matrix is another way to increase the quantum efficiency of their luminescence, since polymer passivates the defects on the QDs' surface and protects against moisture and oxygen, which plays an important role in the degradation of QDs. The functional groups of polymer shell can also be useful in preventing QDs' aggregation and making them hydrophilic, as well as in providing binding sites for bioconjugation and inhibiting non-specific interaction (in particular, to prevent undesirable protein attachment to the QD capping layer). Specifically, polyethylene glycol (PEG), being an inert protective coat for any surface to which it is attached, repelling and being repelled by proteins, is often used as an outer shell of the QDs for biomedical application (Boles et al. 2016).

Moreover, a polymer shell can also be useful in decreasing intrinsic toxicity of the QDs caused by the presence of heavy metal elements in their core. Specifically, the release of Cd^{2+} ions from the QD core in response to ultraviolet (UV) radiation has been assumed to be the main cause of QD cytotoxicity (Derfus et al. 2004; Jamieson et al. 2007; Walling et al. 2009). In particular, during the process of *in vitro* cell imaging under aqueous aerobic conditions, the irradiation of light normally results in the photooxidation of the QDs in live cells (Jin et al. 2015). This leads to electron transfer from the excited QDs to O_2 to produce superoxide O_2^- and an unpaired hole in the QDs. The latter further induces ligand oxidation and cleavage, and the corrosion of the QDs' outer surface generates much more

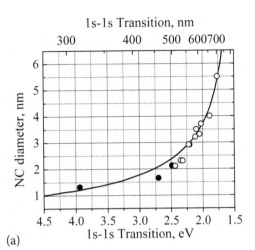

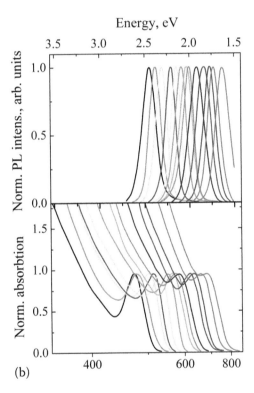

FIGURE 3.1 (a) Sizing curve for thiol-capped CdTe QDs synthesised in water (circles – experimentally determined sizes, solid line – a calculated dependence of 1 s-1 s transition energy on CdTe QD size); (b) Set of typical PL (top) and absorption (down) spectra of TGA-capped and MPA-capped CdTe QDs of different sizes, λ_{exc}=450 nm. (Adapted with permission from Rogach et al. 2007, 14628–37. Copyright 2018, American Chemical Society.)

Cd^{2+} ions (Jin et al. 2015). It has been shown that cadmium ions are able to bind to thiol groups on critical molecules in the mitochondria and produce enough stress and damage to cause cell death (Walling et al. 2009). The Cd^{2+} ions also promote the formation of reactive oxygen species via Cd_2^+-specific cellular pathways. The generated reactive oxygen species can by itself damage the proteins, DNA and lipids, leading to severe cell functional impairments and eventually to cell death (Jin et al. 2015).

To decrease possible toxic effects of the QDs, a new class of Cd-free QDs such as I-III-VI QDs has been developed (Cassette et al. 2013; Zhong et al. 2012). Besides, different polymer materials applied as surface stabilising and coating agents have been proposed in order to improve stability and to decrease toxicity of the QDs (Tomszak et al. 2009; Shen et al. 2011; Matoussi et al. 2012; Jin et al. 2015).

The development of water-soluble highly luminescent QDs covered with appropriate polymer shell opened wide prospects for their application in biosensing and bio-imaging (Zhong et al. 2012; Zhang et al. 2017) (Figure 3.2). For these purposes, a set of the methods for QD conjugation with biomolecules has been developed, including use of electrostatic attraction, covalent-bond formation or streptavidin-biotin linking, silanisation, incorporation of QDs into microbeads and nanobeads, etc. The *in vitro* biomedical and diagnostic applications of the QDs include such techniques as the multicolour fluorescent labelling of cell surface molecules and cellular proteins in microscopy and other applications, detection of pathogens and toxins, DNA and RNA technologies and fluorescence resonance energy transfer (Pelley et al. 2009). QDs are also being explored for use in whole-body *in vivo* imaging of blood vessels, lymph nodes, tissues and tumours, as well for tracking of tumoural, stem or immune cells (Mattoussi et al. 2012; Casette et al. 2013).

On the other hand, simultaneous embedding of QDs of different sizes or composition in a transparent polymer films gives the possibility to create structures, which can emit light

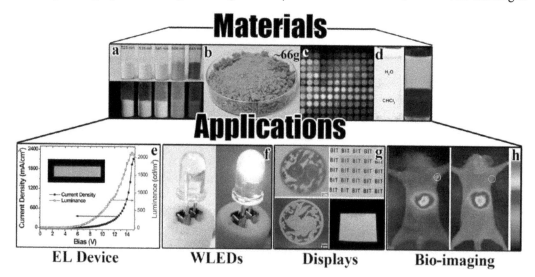

FIGURE 3.2 Highly luminescent and colour-tuneable I-III-VI QDs and the examples of their applications in electroluminescent devices, white LEDs, displays and bio-imaging. (Reprinted with permission from Zhong et al. 2012, 3167–75. Copyright 2018, American Chemical Society.)

in a wide spectral range. These composites became the basis for producing optically excited or electroluminescent white light-emitting diodes (LEDs). For reviews of recent progress in QDs-based white LEDs see (Su et al. 2016, Dai et al. 2017).

In the electroluminescent QD-based LEDs, two strategies are mainly exploited: (i) embedding of QDs into a wide-band gap organic film, where excitons are generated, non-radiatively transferred to the QDs and recombine radiatively, and (ii) sandwiching of a QD layer between electrons and hole transporting layers, providing direct injection of electrons and holes into the QD. Though much work has been done in the optimisation of carrier transport in such structures, including introduction of inorganic or inorganic/organic hybrid matrix, tailoring of the band gap alignment and the use of carrier blocking layers, up to now the reports on QD-based electroluminescent white LEDs are rather limited (Li et al. 2006), but monochromatic QD-based electroluminescent LEDs have already reached the performance characteristics of the state-of-the-art organic LEDs (Dai et al. 2014; Su et al. 2016).

Generally, in phosphor-converted white LEDs, the QDs-polymer composite, which emits light in green and red spectral ranges, is combined with a blue LED as an excitation source, generating a white light with three well-separated RGB peaks. In such devices, QDs are usually blended physically with silicone or epoxy resin directly, followed by thermal curing (Su et al. 2016). Solution processability of QDs and the availability of fabrication techniques to process them into thin films (spin casting, transfer printing, inkjet printing, etc.) are useful to realise white LEDs with large areas and flexibility (Su et al. 2016; Todescatto et al. 2016).

Today, the QD-based phosphor-converted white LEDs are used as a backlight for liquid crystal displays (LCDs). In those devices, the light emitted by white LEDs passes through polarisers, colour filters and liquid crystals, and then the desired colours are obtained and controlled at individual pixel levels (Wang et al. 2017). QDs-based down-conversion displays are now showing performances comparable or even better than organic LED devices. QDs enable highly saturated colours, a wider colour gamut and a comparable response time, while retaining advantages in cost, resolution, optical efficiency and durability (Todescato et al. 2016). QD-based light converters using blue LED chips as back light sources have been applied for display in TVs, monitors and mobile phones. As an example, white LEDs based on multi-shell structured green-emitting CdSe/ZnS/CdSZnS QDs and red-emitting CdSe/CdS/ZnS/CdSZnS QDs with PL QY in the range of 70%–90% under blue LED excitation were successfully integrated into a 46-inch LCD TV panel, demonstrating an excellent colour gamut (Jang et al. 2010). Several display manufacturers (e.g. Samsung, LG, Sony, TCL, etc.) have already introduced QD technology in a global TV market.

It has been demonstrated that embedding QDs in conductive polymers can improve their transport properties (Huynh et al. 2002; Ginger et al. 1999) and increase the efficiency of solar cells based on polymers (Lee et al. 2008). Today, QDs have been extensively considered in PV in the form of the QD Schottky junction solar cell, the QD hetero-junction solar cell, the QD hybrid polymer solar cell, the QD sensitised solar cell, QDs and dye co-sensitised solar cells and the QD rainbow solar cell (Sharma et al. 2016). QD hybrid polymer solar cells are a very promising alternative to conventional crystalline and thin film PV technologies due to their low cost, easy fabrication and high performance. In that system, the QDs are dispersed into a blend of electron and hole-conducting polymers; each type of carrier-transporting polymer has a selective electrical contact to remove the respective charge carriers. The excitons are generated in the QDs under absorption of photons with energy larger or comparable with their band-gap energy. One of the methods applied to enhance efficiency in QD-based PV cells is to create efficient multiple exciton generation in the QDs from a large fraction of the photons in the solar spectrum (Nozik et al. 2010). In fact, strong coulomb coupling between charge carriers was found to generate multi-excitons by a single photon via carrier (or exciton) multiplication (Schaller et al. 2004). In (Ren et al. 2011), the QD-based hybrid solar cells composed of CdS QDs bounded onto crystalline poly (3-hexylthiophene-2, 5-diyl) nanowires through solvent-assisted grafting and ligand exchange have been proposed. In that device, an improved electronic interaction between donor and acceptor components, as well as a well-defined interface in the hybrid photoactive film resulted in enhanced charge separation and transport efficiency of the device. The latter demonstrated an improved maximum power conversion efficiency of 4.1% under AM 1.5 solar illumination.

Another important PV application of QD-polymer composites is in low-cost, solution-processed luminescent solar concentrators (LSCs). LSCSs consist of highly transparent plastic optical waveguides with embedded luminescent species which absorb incident light and emit light at a red-shifted wavelength with high quantum efficiency. The luminescence, guided by total internal reflection, propagates towards a photovoltaic cell placed at the edge of the waveguide, where it is converted into electricity. In Meinardi et al. (2014), the large area LSCs based on the CdSe/CdS quantum dots with giant shells (giant QDs) dispersed in the polymethylmethacrylate (PMMA) matrix have been proposed. These LSCSs demonstrated optical efficiencies >10% and an effective concentration factor of 4.4.

In the next sections, two types of the QD-polymer composites based on water-soluble polymers of gelatine and polyvinyl alcohol (PVA) embedded with II-VI or I-III-VI$_2$ QDs are considered.

3.2 THE METHODS FOR FABRICATION OF THE QD-GELATINE AND QD-PVA COMPOSITES

Generally, the QDs in polymer are produced by the methods of colloidal chemistry. *Colloidal synthesis* of the QDs deals with chemical reactions in solution on a nanometre scale. It has been conducted to make semiconductor nanostructures of different compositions, sizes and shapes. This method involves growing nanoparticles of inorganic materials through chemical reactions of their precursors and, sometimes, controlled precipitation of the reaction product in certain solvents. Generally, the growth process starts with the fast formation of a huge number of nuclei. Then more and more of the solid product deposits onto the nuclei, so the sizes of the crystallite grow slowly until the desired size is reached, at which time the reaction must be quenched. Otherwise, the dots could keep growing under a process known as Oswald ripening, which is the growth of larger dots through the transfer of material from smaller ones, which have a higher solubility. The method needs the usage of stabilisers that hinder coagulation of the particles, preventing their further growth. This property of the stabiliser is determined by the mechanism of inter-action with the surface of the particles. The most frequently used stabilisers are organic compounds such as polyphosphates, trioctylposphines and thiols. The choice of stabiliser is determined by its ability to prevent joining of the particles as well as to provide sufficient passivation of the QD surface. The latter is important for improvement of the intensity of QD emission, since adsorption of molecular groups on the surface of nanocrystal changes its surface potential and decreases the concentration of non-radiative centres.

Two types of polymers acting as stabilisers are often used: gelatine and PVA. These are water-soluble polymers; the former is bio-polymer and the latter is chemically synthesised. To produce QD-polymer composite, two approaches can be used: to grow QDs in polymer solution or to grow QDs in organic solvent or water media and transfer them to the desired polymer.

Gelatine is a readily available, cheap, natural nontoxic and biodegradable polymer (Kozlov et al. 1983; Djagny et al. 2001). Gelatine has long been used in food, pharmaceu-tical and medical industries (Djagny et al. 2001). In the last decades, gelatine nanopar-ticles have been under development as safe and efficient drug- and gene-delivery systems (Djagny et al. 2001; Kaul et al. 2004; Kaul et al. 2005). Gelatine is derived from collagens by partial hydrolysis and denaturation. Collagen is a major protein constituent of connec-tive tissues of vertebrate and invertebrate animals. The defining feature of the collagen molecule is the triple helix, which consists of three parallel left-handed helical polypep-tides coiled about each other to form a right-handed triple helix (Pezron et al. 1990). The primary structure of the collagen chain consists of the amino-acidic sequence of glycine, proline and hydroxyproline. During denaturation, the regular triple helix structure of col-lagen is broken down to form random gelatine coils. Despite the loss of the collagen mac-romolecular organisation, the chemical composition is closely maintained in the resulting gelatine. The elementary unit of gelatine in the isoelectronic state consists of amino (H_2N), carboxyl (COOH), polar and nonpolar basic and acidic groups (R_1) and is represented in the form: H_2N-R_1-COOH. The properties of the gelatine depend on its manufacturing method (acidic or basic), its origin (bovine or pig), the type and number of amino acids and the molecular weight.

Gelatine coils are dissolved in water solution at temperatures of T ~40°C, which is necessary to destroy both hydrogen and electrostatic interactions. When temperature is lowered below roughly 35°C, the gelatine chains undergo a progressive conformational change, known as the coil-to-helix transition. During this process, the solution viscosity is progressively increased so that a transparent, thermo-reversible physical gel is formed. The basic gelation mechanism implies the triple helix renaturation, with the renatured triple helix being the favoured thermodynamic conformation of the gelatine chains below 35°C (Pezron et al. 1990).

Gelatine has been used as a stabilising agent during the synthesis of colloidal CdSe (Xu et al. 2001; Raevskaya et al. 2006), CdS (Skobeeva et al. 2008) and CdTe QDs (Byrne et al. 2007; Parani et al. 2018). In particular, the synthesis of CdSe nanoparticles at the interac-tion between $CdCl_2$ or $CdSO_4$ and Na_2SeSO_3 in aqueous solution of gelatine was proposed in Raevskaya et al. (2006).

In that method, passivation of CdSe QDs with a ZnS shell was achieved by the addition of measured volumes of $Zn(NO_3)_2$ solutions followed by stirring for 5–10 min and the subsequent addition of an equimolar amount of sodium sulphide solution. The aqueous QDs-gelatine solution contained QDs with a molar concentration of 1×10^{-3} M and 10%–20% gelatine. The process of CdS nanocrystal formation due to a reaction between $Cd(NO_3)_2$ and Na_2S in an aqueous solution of gelatine has been described in Skobeeva et al. (2008). Byrne (Byrne et al. 2007) has studied the effect of gelatine on CdTe QD formation. The latter occurred when H_2Te gas was bubbled through a heated aqueous solution of $Cd(ClO_4)_2\cdot6H_2O$ and thioglycolic acid (TGA) stabiliser. The gelatine nanocomposite showed a blue shift in emission wavelengths and stronger emission intensities compared to original TGA-QD samples taken at the same time during the process. It has been proposed that gelatine slows down the Ostwald ripening process and enables the QDs to grow more discretely. The effect of gelatine occurs in a twofold manner. First, the gelatine effectively interacts with and holds the Cd-TGA precursor complex moieties, and thus the distance between these clusters becomes greater. This enables the initial nucleation events to occur on a more individual and separated basis. Second, once formed, the diffusion of these nanoclusters throughout the system during the growth stage is hindered due to increasing viscosity caused by gelatine. Besides the role of the gelatine as a stabiliser, it also acts as a co-capping agent for individual QD stabilisation, which leads to substantial increases in emission intensities when compared to their original counterparts. Kang (Kang et al. 2015) synthesised water-soluble $AgInS_2$/ZnS core-shell QDs in an aqueous solution of gelatine and TGA using an electric pressure cooker as the reaction vessel. The QDs demonstrated a tuneable emission ranging from 535 to 607 nm with a maximum PL QY of up to 39.1%.

The QDs-gelatine solution can be directly used for fabrication of flexible QD-gelatine composite films by the drop-casting technique. In Borkovska et al. (2013b, 2014), thin films of QD-polymer composites were obtained by applying several mL QDs solution onto cleaned glass slides and drying for several days in a black-out drying box at 15°C–20°C and natural ventilation. The thickness of composite films was about 0.18–0.20 mm. Before processing and investigating, the films were removed from the glass. In Borkovska et al. (2015a,b) the aqueous solution of the I-III-VI$_2$ QDs was mixed with 1wt % gelatine, and the resulting viscous solution was drop casted onto glass plates and left for natural drying in the dark at room temperature. In that case, the $CuInS_2$ and $AgInS_2$ NCs were synthesised at room temperature in aqueous media in the presence of mercaptoacetic acid as a stabiliser (Raevskaya et al. 2015; Borkovska et al. 2015a). Generally, the mass fraction of QDs in the composite films does not exceed several % w/v; higher concentrations produce deterioration of optical properties of the solution evidently caused by QD agglomeration and energy transfer between smaller and larger QDs. To increase the concentration of QDs in polymer matrix, special routes are necessary (Ehlert et al. 2015; Smith 2017).

The QD-gelatine composite can be produced also by *encapsulation of QDs into gelatine particles*. In 2000, Coester (Coester et al. 2000) proposed a new two-step desolvation method for manufacturing gelatine nanoparticles with a reduced tendency for aggregation. In this method, after the first desolvation step, the low molecular gelatine fractions present in the supernatant were removed by decanting. The high molecular fractions present in the sediment were redesolved and then desolvated again at pH 2.5 in the second step. The resulting particles were then easily purified by centrifugation and redispersion. The nanoparticles were added after the second desolvation step of the gelatine. After an incubation period of several hours, the particles were cross-linked and purified. In contrast to nanoparticles produced by the one-step desolvation method and nanoparticles based on proteins such as albumin, no sedimentation of the gelatine nanoparticles embedded with QDs was observed during the 3 months' storage period. Following this desolvation method, the possibility to encapsulate CdSe (Chen et al. 2014), CdTe (Wang et al. 2008), PbS (Mozafari et al. 2010) and CdHgTe (Wang et al. 2010) QDs into gelatine nanoparticles has been demonstrated. However, in the case of QDs produced in organic solvents, an additional procedure of ligand exchange to make nanocrystals water-soluble is necessary. For example, to make trioctylphosphine oxide (TOPO)-coated CdSe QDs hydrophilic, Chen (Chen et al. 2014) had realised a ligand exchanging of TOPO with 11-mercaptoundecanoic acid (MUA). After repeating the purification, MUA-modified CdSe QDs were finally redesolved in distilled water and then encapsulated in gelatine nanospheres.

Polyvinyl alcohol is a hydrophilic synthetic linear polymer. It has the idealised formula $[- CH_2 - CH(OH)-]_n$, where n is the degree of polymerisation. The value of n can reach 5000, which means that the PVA molecule can be composed of up to 5000 identical units. Polyvinyl alcohol is produced commercially from polyvinyl acetate, usually by a continuous process. The acetate groups are hydrolysed by ester interchange with methanol in the presence of anhydrous sodium methylate or aqueous sodium hydroxide. The physical characteristics and their specific functional uses depend on the degree of polymerisation and the degree of hydrolysis. In particular, the glass transition and melting temperatures of PVA depend on the degree of hydrolysis. Polyvinyl alcohol is classified into two classes, namely, partially hydrolysed and fully hydrolysed. Partially hydrolysed PVA is used in foods.

PVA is characterised by good thermo-stability, chemical resistance and film forming ability (Pritchard 1970). It is also biodegradable, biocompatible, nontoxic, noncarcinogenic and can be eliminated from the human body by renal excretion. Thus, it is a well-accepted human- and environment-friendly polymer that is used in various pharmaceutical, medical, cosmetic, food and agricultural products (De Merlis et al. 2003). PVA is commonly used in the area of drug delivery as an emulsion stabiliser in the preparation of nanoparticles as well as in coating of particles.

Specifically, PVA is considered a good host material for metal and semiconductor nanoparticles. PVA-protected Ag (Porel et al. 2005), Pt (Luo et al. 2007) and Au (Sun et al. 2009) nanoparticles, as well as QD-PVA composites embedded with ZnO (Sui et al. 2005), PbS (Kuljanin et al. 2006;Pendyala et al. 2009), CdS (Khanna et al. 2005; Raevskaya et al. 2006; Wang et al. 2007; Mansur et al. 2011a, 2012; Rudko et al. 2013), CdSe (Raevskaya et al. 2006; Azizian-Kalandaragh et al. 2010; Borkovska et al. 2013a; Atabey et al. 2013), ZnS (Ozga et al. 2016), ZnCdS (Vineeshkumar et al. 2014), AgInS$_2$ (Borkovska et al. 2016; Wang et al. 2016a), carbon (Wang et al. 2016b), SiC (Saini et al. 2017) and graphene (Kovalchuk et al. 2015; Wang et al. 2018) have been reported. Wang (Wang et al. 2007) had shown that the –OH groups of PVA acted as the coordination sites for cadmium ion aggregations that allowed growing nanosized CdS particles at these sites with the release of S^{2-}ions from thio-acetamide. The effect of cadmium concentration on the in situ synthesis of CdS in PVA has also been studied. Mansur (Mansur et al. 2011b) has synthesised CdS quantum dots (QDs) via aqueous routes using stock solutions of Cd^{2+} and sulphur precursors and acid functionalised PVA (PVA-COOH) or EPC (PVA-C(O)NH-GOx). Mansur (Mansur et al. 2011a) has used carboxylic-functionalised PVA conjugated with bovine serum albumin (BSA) as a capping ligand in the preparation of CdS nanocrystals using cadmium perchlorate and thioacetamide precursors. He has shown that carboxylic-PVA is much more effective on nucleating and stabilising colloidal CdS nanoparticles in aqueous suspensions compared to PVA owing to potentially increased overall interactions. The method of fabricating highly luminescent CdSe/CdS core-shell nanoparticles through a colloidal aqueous route at room temperature using carboxylic-functionalised PVA as a stabilising agent has been proposed (Ramanery et al. 2014). The nanocomposite fibres of uniform and bead-free PVA filled with different loadings of CdSe/ZnS QDs were prepared by the electrospinning process by Atabey et al. (2013).

To produce QD-PVA composite, the QDs synthesised in aqueous solution or gelatine solution can be embedded in a PVA matrix too. In particular, in Borkovska et al. (2013a,b) the CdSe QDs with molar concentrations of 1×10^{-3} M produced in aqueous solutions of gelatine were mixed with 10% PVA and 0.5% gelatine. To improve the homogeneity of the polymer mixture, a hydrochloric acid was also added. In Borkovska et al. (2016), the $CuInS_2$ and $AgInS_2$ QDs synthesised in aqueous solution in the presence of mercapoacids were mixed with a 20 w% aqueous solution of PVA, and the QD-PVA composite films were prepared.

3.3 OPTICAL AND STRUCTURAL PROPERTIES OF THE QD-GELATINE AND QD-PVA COMPOSITES

The X-ray diffraction (XRD) patterns of the composite films of QDs-gelatine and QDs-PVA (Figure 3.3) show several diffraction peaks and halos that are characteristic of crystalline and amorphous phases of conventional semi-crystalline polymers (Yakimets et al. 2005; Pitchard 1970). In particular, the XRD patterns of pristine gelatine and QD-gelatine composite demonstrate the diffuse halo at about and $2\theta = 20°$ of amorphous phase and the peak around $2\theta = 7.8°$, which corresponds to the periodicity from 8.6 to 14.5 Å, characteristic of collagen rod-like triple helices, 300 nm long and 1.5 nm wide, and is usually assigned to the triple-helical crystalline structure in collagen and renatured gelatine (Yakimets et al. 2005). In turn, the XRD patterns of PVA and QD-PVA films show an intense peak around $2\theta = 19.5°$ and the peaks of much lower intensities at $2\theta = 11.5°$, $22.8°$ and $40.7°$, in addition to a diffuse halo at about $2\theta = 21.5°$. In 1948, Bunn had proposed a crystal structure of PVOH that contained a random distribution of syndiotactic and isotactic units (Bunn et al. 1948). The cell is monoclinic (a = 7.81 A, b = 2.52 A, c = 5.51 A, $\alpha = \gamma = 90°$, $\beta = 91.7°$) and each unit cell contains two atactic chains. The polymer chains are described as lying along the b-axis of the unit cell. The most intense peak at $2\theta = 19.5°$ can be ascribed to (10$\bar{1}$) reflection of monoclinic PVA crystal. No noticeable changes are found in the XRD patterns of composites embedded with QDs as compared to corresponding polymers. The peaks from crystalline QD phase are not observed apparently because of a low concentration of QDs in the composites.

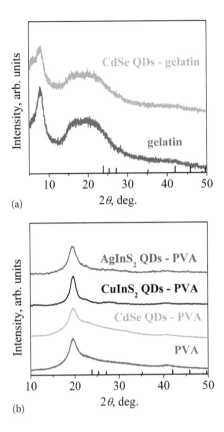

FIGURE 3.3 X-ray diffraction patterns of CdSe QDs-gelatine and pristine gelatine films (a) as well as of CdSe QDs-PVA, AgInS$_2$ QDs-PVA, CuInS$_2$ QDs-PVA and pristine PVA films (b). The line spectrum indicates the reflections for bulk hexagonal CdSe. (Adapted (a) from *Physica B: Cond Mat*, 453, Borkovska et al., Photoluminescence and structural properties of CdSe quantum dot-gelatine composite films, 86–91, Copyright 2018, with permission from Elsevier, and (b) reprinted from Borkovska et al., *AIMS Mat Sci*, 3, 658–68, 2016 with permission under the terms of the Creative Commons Attribution 4.0 International License, open access AIMS.)

The QD-polymer composites were optically clear, and no structural order was observed. This agrees with other reports that the surface of collagen-based films is homogeneous and smooth, and no structure is visible even at high magnifications (Yakimets et al. 2005). The films of pure gelatine or PVA were colourless, while the composite films had yellow or reddish-yellow colour caused by exciton absorption in the QDs. The latter strongly depends on the QD size (Figure 3.4), and the absorption edge shifts to higher energies as the size of the QDs decreases due to the quantum confinement effect (Baskoutas et al. 2006).

The Raman spectra of the QD-polymer composite depend on the excitation wavelength used. The spectra recorded under excitation above the band gap of the QDs show the weak phonon modes of the QD core on a background of the PL signal. Specifically, the Raman spectra of CdSe QDs-gelatine composite show a longitudinal optical (LO) phonon mode of CdSe core at about 205.0–206.5 cm^{-1} (Insert in Figure 3.4a). The peak is shifted towards the low-frequency region as compared to the position of the bulk CdSe peak at 210 cm^{-1} due to the spatial phonon confinement and compressive strains in the QDs (Meulenberg et al. 2004).

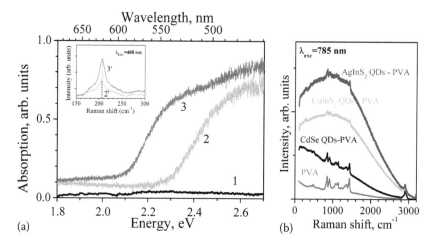

FIGURE 3.4 (a) Optical absorption (1, 2, 3) and Raman spectra (2′, 3′) of pristine gelatine films (1) and CdSe QDs-gelatine composites with effective radius of CdSe QDs of about 1.1 nm (2, 2′) and 1.5 nm (3, 3′); (b) Raman spectra of pristine PVA and QDs-PVA composites under non-resonant excitation. (Adapted (b) from Borkovska et al., *AIMS Mat Sci*, 3, 658–68, 2016 with permission under the terms of the Creative Commons Attribution 4.0 International License, open access AIMS.)

The Raman spectra of the QD-polymer composites under excitation below the band gap of the QDs show mainly the vibrational bands of polymer (Figure 3.4b). Specifically, the vibrational spectra of different QD-PVA composites presented in Figure 3.4b exhibit several strong scattering peaks characteristic of the PVA. The peaks are caused by stretching, bending and wagging vibrations of the O-H, C-H, C-C, C-O and CH_2 groups. The frequency for major Raman bands in the pure PVA and NC-PVA composites and the assignment of the bands are done in Borkovska et al. (2016). The Raman peaks are found on a broad background signal. The nature of this feature is not clear. Incorporation of the QDs into the PVA matrix increases this background. The most intense background was observed for the $AgInS_2$ QDs-PVA composite.

The PL spectra of both CdSe QDs-gelatine and CdSe QDs-PVA composites are dominated by two bands denoted as I_1 and I_2 (Figure 3.5a, b). Usually, the high-energy band (I_1) in the PL spectra of CdSe QDs is ascribed to exciton radiative recombination or radiative recombination of carriers via shallow levels of defects in the QDs (band-edge PL), while the low energy band (I_2) is attributed to radiative recombination of carriers via deep levels of surface defects in the QDs (defect-related PL) (Kortan et al. 1990; Lifshitz et al. 1998). The PL excitation (PLE) spectra detected in the maximum of the I_2 band (Figure 3.5a, b, curves 2) contain a distinct shoulder caused apparently by the light absorption via the ground states in the QDs. The PLE spectra are found to be similar to the optical absorption spectra of the composites (Figure 3.5a, b, curves 2). The PLE spectrum of the I_1 band (not shown) is similar to those of the I_2 band. The Stokes shift, i.e. the energy difference between the PL band position and the shoulder in the PLE spectrum, was of about 0.2 eV and 0.7 eV for the I_1 and I_2 bands, correspondingly.

In the room-temperature PL spectra of the QDs-gelatine composite (Figure 3.5a), the band-edge PL band (I_1) dominates. The PL properties of the QD-PVA composite containing

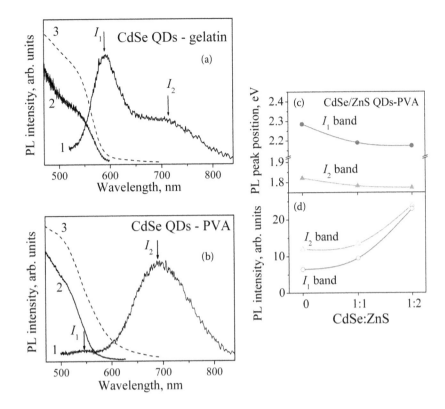

FIGURE 3.5 Room-temperature PL (1), PL excitation (2) and optical absorption (3) spectra of CdSe QDs-gelatine (a) and CdSe QDs-PVA (b) composites. Spectral position (c) and PL intensity (d) of the PL bands in CdSe/ZnS QDs-PVA composite via the ZnS quantity. 2 = 300 K, λ_{exc} = 470 nm. (Adapted (c) from Borkovska et al., *MRS Proceedings*, 1617, 171–7, 2013b with permission. Copyright 2018, Cambridge University Press.)

QDs produced in gelatine and embedded in PVA film without adding HCl are similar to those of the QD-gelatine. It was supposed that some functional groups of gelatine (in particular, amine groups) remain bonded with surface states of the QDs and are not replaced by functional groups of PVA. In contrast, the QDs-PVA composites produced with adding hydrochloric acid were characterised by a lower PL intensity, a lower relative contribution of the band-edge emission to the PL spectrum and lower activation energies of thermal quenching of the PL bands as compared to that of QD-gelatine composite (Figure 3.5b). The similar PL properties (including photo- and thermal stability) were found for the QD-PVA composites containing QDs synthesised in water (Borkovska et al. 2016). Therefore, an insufficient passivation of QD surface defects by functional groups of PVA as compared with that of gelatine has been assumed.

Relative contributions of the I_1 and I_2 bands to the PL spectrum depend not only on the type of polymer matrix but also on the size of CdSe QDs, the presence of shell, as well as on temperature (Figure 3.6). In particular, the passivation of CdSe nanocrystals with ZnS shell results in an increasing of the PL intensity and a shifting of the PL bands to lower energies (Figure 3.5c), as usually observed for CdSe/ZnS QDs

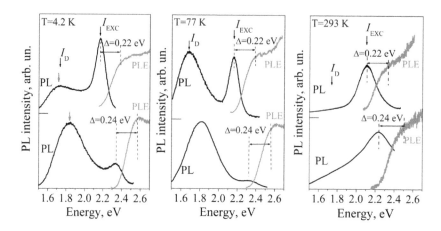

FIGURE 3.6   PL and PLE spectra of the QDs-gelatine composites with effective radius of CdSe QDs of about 1.1 nm (lower curves) and 1.5 nm (upper curves) measured at 4.2 K, 77 K and 293 K, $\lambda_{exc} = 470$ nm.

(Dabbousi et al. 1997; Baranov et al. 2003). The shift is usually accounted for partial tunnelling of the electron wavefunction into the shell (Dabbousi et al. 1997). A large contribution of deep trap emission is found for small QDs (Figure 3.6). This is often observed for bare CdSe QDs of small size passivated by organic capping groups and is explained by large surface-to-volume ratio and insufficient surface defect passivation by organic ligands (Dabbousi et al. 1997). A contribution of the I_1 and I_2 bands to the PL spectrum varies with temperature (Figure 3.6) owing to different thermal quenching of the I_1 and I_2 band intensities (Borkovska et al. 2012). The latter depends also on QD size, polymer type and the presence of inorganic shell.

In contrast to II-VI QDs, the absorption spectra of the $CuInS_2$ and $AgInS_2$ QDs do not exhibit well-defined exciton absorption peaks. This is true for both bare and ZnS-capped QDs dispersed in aqueous buffer solution or embedded in the polymer film (Figure 3.7a) and agrees with the results of other authors (Mao et al. 2011; Hong et al. 2012). The PL spectrum of the I-III-VI$_2$ QDs shows a wide band (or several bands) in the visible spectral range with a large Stokes shift of several hundred meVs (Figure 3.7b). The PLE spectra are similar to optical absorption spectrum. The PL of the I-III-VI$_2$ QDs is commonly ascribed to free-to-bound, bound-to-free or bound-to-bound transitions in donor-acceptor pairs (Mao et al. 2011; Hong et al. 2012; Zhong et al. 2012). The PL band shifts to shorter wavelengths, and the PL intensity increases as the QDs are covered with a ZnS shell or transferred into a gelatine matrix. This can be ascribed to improved passivation of surface defects by the ZnS shell and functional groups of gelatine of the QDs of smaller sizes.

The PL spectra of the QD-polymer composite exhibit not only QD-related PL but also weak emission originated from polymer matrix. For example, the films of both pure gelatine and CdSe QDs-gelatine composite show several overlapped PL bands in the UV, green and yellow spectral ranges caused by functional groups of gelatine (Figure 3.8). The collagen fluorescence in the visible spectral range has been proposed to be of excimer nature and attributed to phenylalanine residues (Volkov et al. 1991). An important feature of

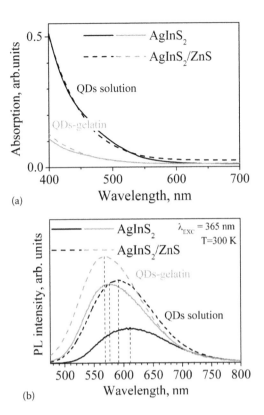

(a)

(b)

FIGURE 3.7 Room temperature optical absorption (a) and PL (b) spectra of the AgInS$_2$ and AgInS$_2$/ZnS QDs dispersed in aqueous solution and embedded in gelatine film (b), $\lambda_{EXC} = 365$ nm. (Adapted from *Mat Sci Semicond Process*, 37, Borkovska, L., Romanyuk, A., Strelchuk, V. et al., Optical characterization of the AgInS$_2$ nanocrystals synthesised in aqueous media under stoichiometric conditions, 135–142, Copyright 2018, with permission from Elsevier.)

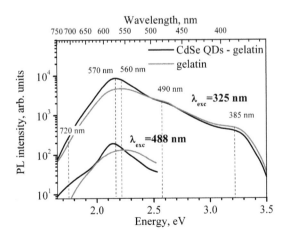

FIGURE 3.8 Room-temperature PL spectra of the CdSe QDs-gelatine and pristine gelatine films under 325 nm and 488 nm laser excitation.

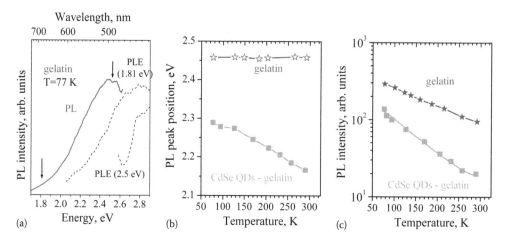

FIGURE 3.9 PL and PLE spectra of pristine gelatine film at $T = 77$ K under $\lambda_{exc} = 409$ nm; the arrows mark detection energies for excitation spectra (a); temperature dependencies of PL peak position (b) and intensity (c) of gelatine and CdSe QDs-gelatine composite, $\lambda_{exc} = 409$ nm.

polymers is the presence of different kinds of radicals and bonds that are not described by its chemical formula. The types and concentrations of these imperfections depend on the synthesis process and processing. Those groups can absorb and emit within the HOMO-LUMO energy gap in the polymer and are called chromophore groups. Generally, this emission is reduced in composites under certain excitations due to the screening effect of the QDs effectively absorbing the excitation light. Alternatively, the quenching of the PL from the polymer matrix in the CdSe QD-polymer composites has been ascribed to rapid charge transfer (Kucur et al. 2004) or efficient Forster resonant energy transfer (Anni et al. 2004) from polymers to the QDs.

If concentration of the QDs in the composite is low, the PL from polymer can be comparable with that of the QDs and wrongly ascribed to QD luminescence. The PLE spectra of polymer emission show a shoulder in the blue-green spectral range and a tail stretched in the red (Figure 3.9a). The intensity of polymer PL increases as the temperature is decreased similarly to QD-related PL (Figure 3.9c). However, a PL peak position does not change with temperature decreasing in contrast to the shifting of the I_1 band to lower energies due to QD band gap shrinkage (Figure 3.9b).

3.4 THERMAL STABILITY OF THE QD-GELATINE AND QD-PVA COMPOSITES

Thermal and photo-stability of QD-polymer composites is an important demand for their applications in light-emitting devices. The effect of thermal annealing of CdSe QD-gelatine composite and pristine gelatine films at temperatures in the range of 100°C–190°C has been studied in Borkovska et al. (2012, 2013b, 2014).

The annealing of the CdSe QD-gelatine composite at temperatures in the range of 100°C –130°C resulted in the shifting of both band-edge (I_1) and defect-related (I_2) PL bands of QDs to the low-energy region (red shift effect), and the changing in the PL intensity (Figure 3.10a). The red shift was observed at 300 and 77 K and was somewhat larger for the

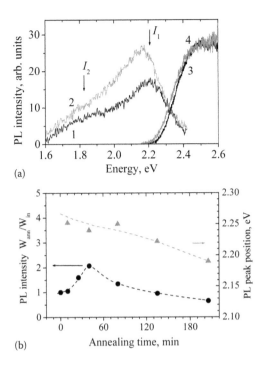

FIGURE 3.10   Room-temperature PL (1, 2) and PLE spectra of I_2 band (3, 4) for CdSe QDs-gelatine composite before (1, 3) and after (2, 4) thermal annealing at 100°C for 1 h, λ_{exc} = 470 nm (a). Relative change of the intensity (circles) and spectral position (triangles) of the I_1 band versus annealing time for CdSe QDs-gelatine composite (b), dashed lines are drawn for a better eye guide. (Adapted from Borkovska et al. (2013b) with permission. Copyright 2018, Cambridge University Press.)

I_1 band. In different samples, the shift magnitude ranges between 30 and 60 meV. In the PLE spectra, a comparable red shift of the QD-related peak was also observed. The PL intensity recorded at 77 K decreased under annealing, and the room-temperature PL increased as a rule. As the annealing time increased, the PL intensity increased at first (up to 1 hour of annealing) and then dropped down, while the spectral position of the I_1 band progressively shifted to lower energies (Figure 3.10b). In some CdSe QDs-gelatine composites, the room-temperature PL intensity decreased even under thermal annealing at 100°C for 1 h and continued decreasing as the annealing temperature increased up to 130°C.

Analysis of the temperature dependence of the I_1 band intensity in the range of 77–390 K showed that corresponding Arrhenius plots can be approximated by two linear dependencies in the ranges of 77–200 and 250–390 K, respectively (Figure 3.11). The fitting of the temperature dependencies of the PL intensity $W(T)$ by two exponents:

$$W(T) = W_0 \left(1 + C_1 \exp(-E_1 / kT) + C_2 \exp(-E_2 / kT)\right)^{-1} \tag{3.1}$$

showed that thermal annealing at 100°C–130°C resulted in the decrease of both activation energies of the PL thermal quenching (Table 3.1).

The absorption spectra of the QD-gelatine composites did not change upon the annealing at 100°C. The same thermal treatment of gelatine films also didn't change film

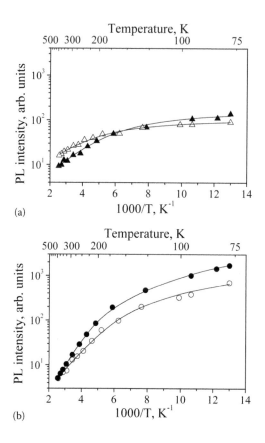

(a)

(b)

FIGURE 3.11   PL intensity for the I_1 (a) and I_2 bands (b) measured before (solid symbols) and after (open symbols) thermal annealing at 100°C for 1 h as a function of the sample temperature. $\lambda_{exc} = 470$ nm. (Adapted from Borkovska et al. (2012) with permission. Copyright 2018, John Wiley and Sons.)

TABLE 3.1   The Activation Energies of Thermal Quenching of the I_1 and I_2 Band Intensities in the Initial and Annealed at 100°C for 1 h Films of CdSe QDs-gelatine Composite, $\lambda_{exc} = 470$ nm

The PL band	E_1, meV (77–150 K)		E_2, meV (200–390 K)	
	Initial	Annealed	Initial	Annealed
I_1 band	11.5	9.5	62	57
I_2 band	26.1	18.7	123	87

Source: Reprinted from Borkovska et al., *Phys Stat Sol C*, 9, 1779–82, 2012 with permission. Copyright 2018, John Wiley and Sons.

transparency in the visible spectral range but resulted in the increasing in the intensity of gelatine PL.

The effects stimulated by thermal treatment at 100°C were found to be partially reversible. The storage of the annealed samples at room ambience for about a week resulted in full or partial recovery of the initial PL characteristics. For small PL changes produced by the annealing, the PL spectra recovered completely. The passivation of CdSe QDs with a ZnS shell decreased the effects of annealing; the higher the ZnS volume, the lower the effects.

The annealing at temperatures above 130°C produced different changes in the optical properties of composites (Borkovska et al. 2014). In particular, the annealing of the QD-gelatine composite at 140°C stimulated the increase in the intensity of the band-edge PL both at 77 and 300 K. Similar annealing of gelatine films produced the increasing of the PL intensity and shifting of the absorption edge of polymer to the green. At higher annealing temperatures (≥160°C), a pronounced degradation of the PL properties of the composite and gelatine films was found. Both types of films became brown and extremely brittle. Moreover, certain signs of film melting were found. However, storage for six months at ambient conditions of the composite annealed at 190°C resulted in a pronounced increase of the PL intensity originated from the gelatine matrix; the PL intensity enhanced that of the untreated film.

The XRD study of the annealed composite and gelatine films revealed the decreasing in the intensity of the peak $2\theta = 7.8$ grad, indicating the helix-to-coil conformational transformation of solid gelatine macromolecules. A pronounced decrease was observed even for the films annealed at 100°C. This phenomenon is accompanied by a release of water molecules, which form hydrogen-bonded bridges, stabilising the helix (Kozlov et al. 1983), and it agrees with the reduction of film flexibility due to annealing. As a result of this transition, some functional groups of gelatine become detached from one another as well as from the QD surface. Therefore, it has been supposed that the changes in QD passivating ligands resulting from structural transformations in the gelatine matrix are responsible mainly for the red shift of the QD PL and the changes in the PL intensity observed upon the annealing at 100°C. Specifically, gelatine can passivate surface Cd atoms with amino- and carboxy-groups, as well as with -NH-CO- fragments of the polymer chains. Thermal annealing can cause a dissociation of some of these coordination bonds, for example, the ones with amino-groups. Recovery of the initial PL characteristics can be explained by re-bonding of Cd surface atoms with functional groups of gelatine.

The decrease of the PL intensity of the composites subjected to thermal annealing at 160°C can be the result of strongly increased visible light absorption in the gelatine matrix due to the appearance of chromophore groups. The annealing of gelatine at elevated temperatures stimulates the breaking of hydrogen bonds between the macromolecules as well as between neighbouring parts of a macromolecule, the splitting of polymer chains and the formation of free radicals. The transformation of free radicals, including chain-transfer reactions between free macro-radicals, can give rise to a partial cross-linking of polymer chains and the formation of carbonyl ($C=O$) and carboxyl (–COOH) groups as well as of polyene units, which are known to be quite common absorbing and light-emitting species in polymers (Rudko et al. 2012). In fact, in Yannas (1967), an insolubilisation of thermally treated gelatine had been ascribed to excessive dehydration, resulting in the formation of a three-dimensional network by the interchain covalent cross-linking. The thermal cross-linking mechanism was proposed to be a condensation reaction between a carboxyl group and an amine group of adjacent gelatine chains with an excessive dehydration. It can be supposed that some of these processes promote the passivation of surface defects in the QDs and are responsible for increasing of the PL intensity in composite annealed at 140°C. The above-mentioned transformations can occur both under annealing at elevated temperatures and during a storage of annealed polymer at room temperature. The latter can explain the increase of the PL intensity in the annealed films during their storage in the air for months.

The effect of thermal annealing on the PL spectra of the QDs-PVA composites was somewhat different from what was observed in the QDs-gelatine. In the PL and PLE spectra of CdSe QDs-PVA composites thermally treated at 90–120°C, no spectral shifts were found. At the same time, the PL intensity increases both at 77 and 300 K (Figure 3.12a); the higher the annealing temperature, the more the PL rose (Figure 3.12a). Moreover, in contrast to the QDs-gelatine composite, the PL intensity of QDs-PVA films increased and then saturated as the duration of annealing increased. Though the PL bands in the QDs-PVA composite did not shift during the first hour of annealing, a small red shift of the PL bands was found for long annealing times. Passivation of CdSe QDs with a ZnS shell decreased the effect of annealing on the QD PL spectra, and the higher the [ZnS]:[CdSe] ratio, the lower the effect (Borkovska et al. 2013b). The effect of the annealing on the PL spectra of the QDs-PVA composites was found to be irreversible.

In the XRD patterns, thermal annealing of pristine PVA and QD-PVA composite films increased the intensity of the PVA-related peaks, indicating the increase of the degree of crystallinity of the PVA matrix. It was found that the higher the annealing temperature, the more pronounced were the peaks of crystalline PVA (Figure 3.13a). This was in agreement with the results of Raman spectra investigations. In particular, the intensities of the sharp Raman peak at 1146 cm^{-1} and the less pronounced band at about 1121 cm^{-1} (Figure 3.13b) are used as an indicator of crystallinity and "amorphicity" of the polymer, respectively (Iwamoto et al. 1979). The ratio of the intensities of these bands is typical for partially crystallised PVA. The annealing decreased the relative intensity of the band at about 1121 cm^{-1} (Figure 3.13b), indicating further crystallisation of the PVA matrix.

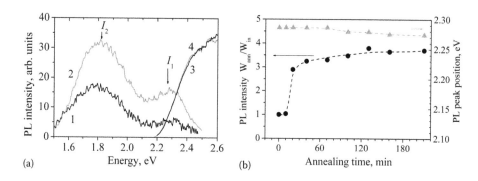

FIGURE 3.12 Room-temperature PL (1, 2) and PLE spectra of I_2 band (3, 4) for CdSe QDs-PVA composite before (1, 3) and after (2, 4) thermal annealing at 100°C for 1 h, $\lambda_{exc} = 470$ nm (a). Relative change of the intensity (circles) and spectral position (triangles) of the I_1 band versus annealing time for CdSe QDs-PVA composite (b), dashed lines are drawn for a better eye guide. (Adapted from Borkovska et al. (2013b) with permission. Copyright 2018, Cambridge University Press.)

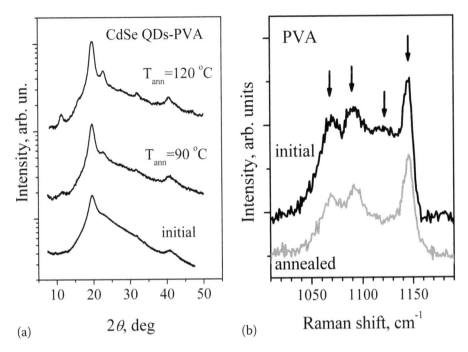

FIGURE 3.13 (a) X-ray diffraction patterns of the CdSe QDs-PVA composite before and after annealing at 90°C and 120°C (a), the curves are shifted in vertical direction for clarity and (b) the Raman spectra of the PVA films before and after annealing at 100°C (b), $\lambda_{exc} = 785$ nm. (Adapted (a) from *Appl Surf Sci*, 281, Borkovska et al., Enhancement of the photoluminescence in CdSe quantum dot-polyvinyl alcohol composite by light irradiation, 118–22, with permission, copyright 2018, Elsevier, and (b) reprinted with permission from Borkovska et al. (2016) under the terms of the Creative Commons Attribution 4.0 International License, open access AIMS.)

The effect of thermal annealing at 100°C on the PL spectra of the QD-PVA was ascribed to enhanced passivation of non-radiative defects on the QD surface stimulated by rearrangement of some functional groups or fragments of the polymer chains of PVA matrix, which occurred during oxidation of the PVA matrix.

3.5 PHOTO-STABILITY OF THE QD-PVA AND QD-GELATINE COMPOSITES

The applications of the QDs-polymer composites in the light-emitting devices put forward high requirements for their photo-stability. This is especially important in the case of PVA, which is known to undergo photochemical transformations under γ-quanta or UV-light irradiation (Etsuo et al. 1978). The introduction of QDs into the PVA matrix strongly influences the photo-stability of polymer. The photo-stability of the pristine PVA and QD-PVA composites has been studied in Bol et al. (2001); Pendyala et al. (2009); Rudko et al. (2012); Borkovska et al. (2013); Rudko et al. (2015); and Borkovska et al. (2016). In particular, Rudko has demonstrated that UV-induced changes in the PL of the PVA matrix in CdS QDs-PVA composite depend on both the excitation power of the 266 nm laser beam and the duration of irradiation (Rudko et al. 2012). The UV-induced changes in the nanocomposite were shown to be predominantly governed by the processes at the QD/polymer interface. An improvement of PVA stability against UV illumination achieved by the introduction of CdS QDs into polymer has been reported in Rudko et al. (2015). The effect was ascribed to diminishing the probability of photo-activated formation of defects in polymer due to the lowering of the efficiency of polymer excitation via partial absorption of incident light by the embedded QDs, as well as the de-excitation of the macromolecules that have already absorbed UV quanta via energy drain to nanoparticles.

The effect of photo-induced enhancement (more than tenfold) of room temperature deep trap emission of CdSe QDs embedded in PVA film has been discovered in Borkovska et al. (2013a). The effect was observed upon illumination of the QD-PVA composite by LED light of 409 or 470 nm at elevated temperatures. In particular, irradiation of the QD-PVA composite during short term annealing at 90°C noticeably enhanced the effect of thermal treatment, i.e. it resulted in a stronger increase of the PL intensity (Figure 3.14a). The effect of photo-induced PL enhancement started already at 300 K (inset in the Figure 3.14b) and decreased at 120°C. The effect was shown to be caused by an increase of the activation energy of the thermal quenching of defect-related PL in 1.5–2 times (Figure 3.14b). It has been supposed that photocarriers generated in the QDs promote the formation of free radicals and their secondary reactions in the PVA matrix, most likely near the QD/PVA interface. The photo-induced transformations of PVA polymer chains stimulate reconstruction of passivating ligands at the QD/PVA interface and change the energy position of the levels of non-radiative defects.

The photo-brightening effect of the PbS QDs-PVA composite has been reported in Pendyala et al. (2009), and of ZnS:Mn²⁺ QDs-PVA in Bol et al. (2001). The effect consisted

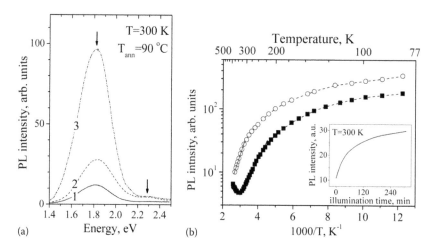

FIGURE 3.14 (a) Room-temperature PL spectra of the CdSe QDs-PVA composite before (1) and after annealing at 90°C in the dark (2) and under illumination by LED light of 409 nm (3), $\lambda_{exc} = 470$ nm. (b) Temperature dependence of I_2 band intensity before (solid symbols) and after (open symbols) thermal annealing at 90°C upon illumination by LED light and their approximation by single-exponential dependence (dashed lines); inset: evolution of the defect-related PL band intensity upon illumination by LED light of 470 nm at 300 K. (Adapted with permission from *Appl Surf Sci*, 281, Borkovska et al., Enhancement of the photoluminescence in CdSe quantum dot-polyvinyl alcohol composite by light irradiation, 118–22. Copyright 2018, Elsevier.)

in the increasing of the QD PL intensity under prolonged light irradiation and ascribed to the curing of the polymer with laser. It has been found that the presence of excess ions of other metals at the surface of QDs changes their property to photo-darkening/brightening, which depends on the direction of carrier transfer mechanism (from QDs to the surface-adsorbed metal ions or vice-versa).

The effect of LED light irradiation on the PL intensity of the $AgInS_2$ QDs-PVA composite has been reported in Borkovska et al. (2016). Specifically, an irradiation of the composite by LED light during thermal treatment decreased the effect of annealing (i.e. resulted in smaller enhancement of the QD PL intensity) as well as produced a pronounced darkening of the composite (Figure 3.15a). The same photo-darkening effect occurred in the composite at room temperature but needed more time (Figure 3.15c). The colour and PL intensity of the $AgInS_2$ QDs-PVA composite was restored during storage of the films at ambient conditions for several weeks. The effect of irradiation was ascribed to photo-stimulated formation of some light-absorbing species from the residual Ag-related compounds (presumably, the silver nanoparticles or clusters) near the QD surface, while restoration of the optical properties of the composite was ascribed to oxidation of these silver-related species.

It should be noted that the $AgInS_2$ QDs dispersed in water solution demonstrate the same photo-darkening effects, i.e. the decreasing of the QD PL intensity and darkening of the solution under irradiation by a UV light (Figure 3.15b). At the same time, adding a small amount of gelatine to the solution not only hinders these processes but also results in an increase of the PL intensity. This can be ascribed to improved passivation of QD surface states by functional groups of gelatine, in particular by amino-groups. In fact, in Joo et al. (2018), the increase of the QY% of the as-synthesised CdSe QDs by the amino passivation from 37% to 75% and the QY% of the CdSe/ZnSe core-shell QDs by 28% has been reported. The photo-darkening effects (the decreasing of PL intensity and darkening of the composite) were not observed for $CuInS_2$ QDs-PVA composite nor for both $AgInS_2$ and $CuInS_2$ QDs embedded in gelatine matrix, indicating better photo-stability of the QD-gelatine composite as compared to the QD-PVA.

3.6 APPLICATIONS OF THE QD-GELATINE AND QD-PVA COMPOSITES IN BIO-IMAGING, SENSING AND LEDS

The QD-composite to be applied for *in vivo* bio-imaging should be not only water-soluble but also biocompatible and nontoxic. Both gelatine and PVA nanocomposites are well satisfying these demands (Ulubayram et al. 2002; DeMerlis et al. 2003; Jin et al. 2015). Natural proteins are being widely used as potential carriers for site-specific drug delivery, as they demonstrate nontoxic and biocompatible behaviour. At present, gelatine nanocomposites play a crucial role in various aspects of biology and medical sciences, in particular, in delivery of drugs and genes, as well as for *in vivo* pharmacological performances. Various cross-linkers used to improve the physicochemical behaviour of gelatine nanocomposites have been reported.

Byrne and colleagues (Byrne et al. 2007) have demonstrated that gelatine-QD composites can readily pass through the cell membrane and illuminate the cytoskeleton of the THP-1 macrophage cells. The cellular toxicity response has been measured using a

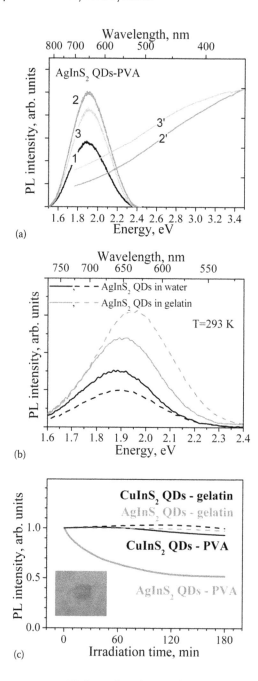

FIGURE 3.15 (a) Room-temperature PL (1, 2, 3) and optical absorption (2′, 3′) spectra of the AgInS$_2$ QDs-PVA composite before (1) and after annealing at 100°C in dark (2′, 2) and under irradiation with LED light of 409 nm (3, 3′); (b) PL spectra of the AgInS$_2$ QDs dispersed in water and gelatine solution before (solid curves) and after irradiation with 365 nm line of Hg- lamp (dashed curves); (c) evolution of the PL intensity of the AgInS$_2$ QDs-PVA composite upon illumination by LED light of 409 nm at 293 K. (Adapted from Borkovska et al., *AIMS Mat Sci*, 3, 658–68, 2016. with permission under the terms of the Creative Commons Attribution 4.0 International License, open access AIMS.)

multiparameter cytotoxicity, where dye luminescence intensity is correlated to toxic phenomena. In comparison to the original TGA QDs, the gelatine-QDs displayed much lower rates of cytotoxicity and improved biocompatibility. Aswathy et al. (2011) have exploited CdTe/CdSe QDs' embedded gelatine nanoparticles for *in vitro* cellular imaging. The gelatine-coated QDs were found to be highly biocompatible in different cell lines, including GI-1, MCF 7, HUVEC and L929 even at higher concentrations of nanoparticles. Murata et al. (2017) have prepared a multimodal probe which can simultaneously visualise cells by optical and magnetic resonance imaging modalities. The probe utilised composites of gelatine nanospheres, incorporating both QDs and iron oxide nanoparticles. These composites incubated with normal human articular chondrocytes were efficiently internalised into the cells, although their cytotoxicity was observed at the R8 concentration of 320 μM. Kang et al. (2015) have shown that gelatine-coated $AgInS_2$/ZnS core-shell QDs possessed excellent photo-stability and water/buffer stability, as well as negligible toxicity. In particular, these QDs showed intense intracellular luminescence upon incubation with HeLa cells, providing the function of cell imaging, as well as showing no obvious cytotoxic effect on HeLa cells at 1.5265–100 μg/mL.

The cytotoxicity of CdTe QD-polymer composites was systematically investigated in Jin et al. (2015) using the Cal27 and HeLa cell lines for three kinds of conventional low-cost polymers, namely, PVA, sodium polystyrene sulphonate (PSS) and polydiallyl dimethyl ammonium chloride (PDDA). The experimental results showed that the cytotoxicity decreased in the order CdTe-PDDA>CdTe>CdTe-PSS>CdTe-PVA, indicating that PVA can reduce the toxicity of the QDs by protecting them from photooxidation. In fact, the results of X-ray photoelectron spectroscopy investigations proved that QDs in CdTe-PVA composite were not oxidised after photoexcitation. The CdTe QD-PVA composite showed low cytotoxicity for one week. At least, in Jin et al. (2015), the bio-imaging of Cal27 and HeLa cells was performed using CdTe QD-PSS and CdTe QD-PVA composites with different emission colours, indicating the potential application in simultaneously observing different biological compartments using a single-excitation light source.

Mansur et al. (2012) had shown a successive application of the carboxylic-functionalised PVA-QD system to the bioconjugation of BSA via carboxylic-amine reactions. In Mansur et al. (2011b), a successive application of a hybrid composed of carboxylic-functionalised PVA coupled with glucose oxidase and bioconjugated with CdS QDs as a nanostructured sensor for glucose detecting has been demonstrated. The hybrid was biochemically assayed by injecting the glucose substrate and detecting the formation of peroxide with the enzyme horseradish peroxidase. The hybrid had proven to be simultaneously active as a fluorophore evaluated by PL and by biochemical enzymatic assay produced via one-step aqueous colloidal route and had demonstrated great potential in the development of sensors for targeting peptides, proteins, enzymes and antibodies.

In addition, the QDs-PVA composites have been proposed for applications in sensing. Pendyala et al. (2009) proposed PbS QDs capped with mercaptoethanol (C_2H_5OSH) and synthesised in PVA for PL sensing of various ions such as Zn, Cd, Hg, Ag, Cu, Fe, Mn, Co, Cr and Ni. Choudhury et al. (2010) had proposed ZnO QDs-PVA composites for acetone sensing, which demonstrated high sensitivity and low response times.

The QD-gelatine and QD-PVA composites have been successfully applied in white LEDs too. Cosgun et al. (2015) synthesised cadmium-free and water-soluble ZnSe:Mn/ZnS QDs through a nucleation doping strategy and then embedded them in PVA. The flexible composite contains well-dispersed QDs and exhibits highly efficient PL at 590–635 nm. It was used for the fabrication of resin-free white LEDs, which demonstrated high luminous efficiency, good stability and colour rendering index (CRI) value of 93.5 and correlated colour temperature (CCT) of 2913 K at Commission Internationale de l'Eclairage (CIE) colour coordinates of (0.41, 0.37).

In Liang et al. (2014), highly luminescent and flexible films fabricated via self-assembly of triple building blocks – layered double hydroxide (LDH) nanoplatelets, PVA and QDs (CdTe or CdSe/ZnS), which showed 2D ordered structure and finely tuneable fluorescence (green, yellow, orange and red) – were proposed for application in white LEDs. The LDH/PVA-QDs films displayed high photo- and thermo-stability. The red-emitting LDH/(PVA-CdSe/ZnS) film was incorporated in the commercialised white LED to improve colour rendering properties through modifying its spectral distribution.

Kang et al. (2015) had proposed the method for fabrication of warm white LEDs with gelatine-coated $AgInS_2$/ZnS core-shell QDs synthesised via a facile aqueous method. The gelatine solution containing yellow-, orange- and orange-red-emitting QDs was directly dropped onto the blue (460 nm) LED chip to form a homogeneous QD-gelatine hybrid film after drying. Because of uniform dispersion of QDs on top of the LED, the device shows uniform colour distribution, which is more advantageous than LEDs using bulk phosphor/resin as a wavelength converter. The as-fabricated warm white LED exhibited a luminous efficacy of 39.85 lm/W, a CCT of 2634 K and a CRI of 71 at a drive current of 20 mA.

In Kim et al. (2014), a method for fabrication of free-standing composite film of PVA embedded with green-to-greenish-yellow alloyed Cu-In-Ga-S (CIGS)/ZnS and red InP/ZnS QDs with PL QYs of 70%–73% and 67%, respectively, has been proposed for use as down-converters, in conjunction with a blue LED. The best PL properties demonstrated the structure of red InP/ZnS QD-embedding polyvinylpyrrolidone sequentially stacked onto the $CI_{0.2}G_{0.8}$ S/ZnS QD-PVA film. This dual colour-emitting, flexible and transparent bilayer composite film was integrated in tri-coloured white QD-LED which possessed an exceptional colour rendering property through reinforcing a red spectral component and balancing a white spectral distribution.

Wang et al. (2016a) fabricated luminescent and flexible films of green- and red-emitting $AgInS_2$/ZnS core-shell QDs embedded in PVA that showed a high PL QY of about 55% and 64% upon 460 nm excitation, respectively. Finally, the green and red QD/PVA films were successfully applied on top of a conventional blue InGaN chip for remote-type, warm-white LEDs. As-fabricated, warm white LEDs exhibited a CRI of about 90.2 and a CCT of 3698 K.

Wang et al. (2016a) demonstrated white LEDs based on the PVA films embedded with nitrogen and sulphur co-doped carbon dots and excited with a 365 nm UV-LED chip. The LED emitted warm white light with a colour coordinate of (0.40, 0.38), a CCT of 3980 K and luminance of 3629 cd m^{-2}. Wang et al. (2018) fabricated free-standing luminescent composite films by combining S, N-co-doped graphene QDs, which demonstrated intense blue PL under UV excitation (QY of 51.2%), with yellow-emitting YAG:Ce^{3+} phosphors dispersed in a PVA matrix. The resultant transparent films were easily scalable, fully flexible, highly photo- and thermostable and retained their optical properties in the solid forms. The films were placed on the light concentrator and excited with an array of custom-made 360 nm LEDs. The device showed white-light emission with CIE colour coordinates of (0.296, 0.311), CCT of 7342 K, CRI of 74.6 and LE of 19.9 lm/W, respectively, and demonstrated a high potential for application of fabricated composite in high-powered UV-pumped remote white LEDs.

3.7 CONCLUSIONS

In conclusion, the QD-polymer composites possess strong potentials in biosensing and bio-imaging, providing improvements in detection sensitivity, assay simplicity and low-cost methods. However, the potential toxicity of these nanocomposites, as well as insufficient stability of their QD assays over a period of time under ambient conditions, significantly limits their range of applications. Their performances can be enhanced by improving the methods for synthesis of nontoxic QDs with high biocompatibility and narrow size/shape/conjugation distributions, as well as by improving polymer shell performance to prevent toxic effect and non-specific interactions. The QD-polymer composites also hold high potential for development of cost-effective, large-area, ultrathin and flexible light-emitting and PV devices. For large-scale application, the following aspects still remain critical. First, the technology of QD synthesis needs to be further improved. It concerns the optimisation of passivation processes with the inorganic shell and ligand layer to decrease the concentration of defects that cause poor stability and device degradation as well as the improvement of synthesis procedures for cadmium-free QDs in order to increase their PL QY and decrease the half-width of the PL band. Attention should also be paid to the development of "green", non-expensive methods of QD synthesis compatible with large-scale production. Second, the exploration of new polymer matrices with good compatibility with QDs and the desired electrical and optical properties is still important. More focus should be given to the research of different film fabrication processes, including transfer printing, ink jet printing, patterning, etc. Detailed investigations focusing on degradation mechanisms of QDs-based light-emitting and PV devices, including their thermo- and photo-stability, are still to be fully addressed. Despite these challenges, the QDs-based composites are expected to find their industrial applications in lighting, photovoltaics and biomedical research within the next decade.

REFERENCES

Anni, M., Manna, L. and Cingolani, R. 2004. Förster energy transfer from blue-emitting polymers to colloidal CdSe/ZnS core shell quantum dots. *Appl Phys Lett* 85:4169–71.

Aswathy, R. G., Sivakumar, B., Brahatheeshwaran, D., Ukai, T., Yoshida, Y., Maekawa, T. and Kumar, S. D. 2011. Biocompatible fluorescent jelly quantum dots for bioimaging. *Mater Express* 1:291–8.

Atabey, E., Wei, S., Zhang, X., Gu, H., Yan, X., Huang, Y., Shao, L., He, Q., Zhu, J., Sun, L., Kucknoor, A. S., Wang, A. and Guo, Z. 2013. Fluorescent electrospun polyvinyl alcohol/CdSe@ZnS nanocomposite fibers. *J Composite Materials* 47:3175–85.

Azizian-Kalandaragh, Y. and Khodayari, A. 2010. Ultrasound-assisted preparation of CdSe nanocrystals in the presence of Polyvinyl alcohol as a capping agent. *Mat Sci Semicond Proc* 13:225–230.

Baranov, A. V., Rakovich, Y. P., Donegan, J. F., Perova, T. S., Moore, R. A., Talapin, D. V., Rogach, A. L., Masumoto, Y. and Nabiev, I. 2003. Effect of ZnS shell thickness on the phonon spectra in CdSe quantum dots. *Phys Rev B* 68:165306.

Baskoutas, S. and Terzis, A. F. 2006. Size-dependent band gap of colloidal quantum dots. *J Appl Phys* 99:013708.

Bol, A. A. and Meijerink, A. 2001. Luminescence quantum efficiency of nanocrystalline ZnS:Mn^{2+}. 2. Enhancement by UV irradiation. *J Phys Chem* B105:10203–9.

Boles, M. A., Ling, D., Hyeon, T. and Talapin, D. V. 2016. The surface science of nanocrystals. *Nature Mater* 15:141–53.

Borkovska, L. V., Korsunska, N. O., Stara, T. R., Dzhagan, V. M., Stroyuk, O. L., Raevska, O. Y. and Kryshtab, T. G. 2012. Modification by thermal annealing of the luminescent characteristics of CdSe quantum dots in gelatin films. *Phys Stat Sol C* 9:1779–82.

Borkovska, L., Korsunska, N., Stara, T., Gudymenko, O., Venger, Y., Stroyuk, O., Raevska, O. and Kryshtab, T. 2013a. Enhancement of the photoluminescence in CdSe quantum dot-polyvinyl alcohol composite by light irradiation. *Appl Surf Sci* 281:118–22.

Borkovska, L., Korsunska, N., Stara, T., Bondarenko, V., Gudymenko, O., Stroyuk, O., Raevska, O. and Kryshtab, T. 2013b. Photoluminescence and structural properties of CdSe quantum dot-polymer composite films. *MRS Proceedings* 1617:171–7.

Borkovska, L., Korsunska, N., Stara, T., Gudymenko, O., Kladko, V., Stroyuk, O., Raevskaya, A. and Kryshtab, T. 2014. Photoluminescence and structural properties of CdSe quantum dot-gelatin composite films. *Physica B: Cond Mat* 453:86–91.

Borkovska, L., Romanyuk, A., Strelchuk, V., Polishchuk, Y., Kladko, V., Stroyuk, O., Raevskaya, A. and Kryshtab, T. 2015a. The photoluminescence properties of CuInS$_2$ and AgInS$_2$ nanocrystals synthesized in aqueous solutions. *ECS Trans* 66:171–9.

Borkovska, L., Romanyuk, A., Strelchuk, V., Polishchuk, Y., Kladko, V., Raevskaya, A., Stroyuk, O. and Kryshtab, T. 2015b. Optical characterization of the AgInS$_2$ nanocrystals synthesized in aqueous media under stoichiometric conditions. *Mat Sci Semicond Process* 37:135–42.

Bunn, C. W. 1948. Crystal structure of polyvinyl alcohol. *Nature (London)* 161:929–30.

Byrne, S. J., Williams, Y., Davies, A., Corr, S. A., Rakovich, A., Gun'ko, Y. K., Rakovich, Y. P., Donegan, J. F. and Volkov, Y. 2007. "Jelly dots": Synthesis and cytotoxicity studies of CdTe quantum dot-gelatin nanocomposites. *Small* 3:1152–6.

Cassette, E., Helle, M., Bezdetnaya, L. F., Marchal, F., Dubertret, B. and Pons, T. 2013. Design of new quantum dot materials for deep tissue infrared imaging. *Adv Drug Del Rev* 65:719–31.

Chen, L., Willoughby, A. and Zhang, J. 2014. Luminescent gelatin nanospheres by encapsulating CdSe quantum dots. *Luminescence* 29:74–8.

Choudhury, M., Nath, S. S., Chakdar, D., Gope, G. and Nath, R. J. 2010. Acetone sensing of ZnO quantum dots embedded in polyvinyl alcohol matrix. *Adv Sci Lett* 3:6–9.

Cingarapu, S., Yang, Z., Sorensen, C. M. and Klabunde, K. J. 2012. Synthesis of CdSe/ZnS and CdTe/ZnS Quantum Dots: Refined Digestive Ripening. *J Nanomater* 2012:Article ID 312087.

Coester, C. J., Langer, K., Brisen, H. V. and Kruter, J. 2000. Gelatin nanoparticles by two step desolvation – a new preparation method, surface modifications and cell uptake. *J Microencapsul* 17:187–93.

Cosgun, A., Fu, R., Jiang, W., Li, J., Song, J., Song, X. and Zeng, H. 2015. Flexible quantum dot-PVA composites for white LEDs. *J Mater Chem C* 3:257–64.

Crooker, A., Barrich, R., Hollingsworth, J. A. and Klimov, V. I. 2003. Multiple temperature regimes of radiative decay in CdSe nanocrystal quantum dots: Intrinsic limits to the dark-exciton lifetime. *Appl Phys Lett* 82:2793.

Dabbousi, B. O., Rodriguez-Viejo, J., Mikulec, F. V., Heine, J. R., Mattoussi, H., Ober, R., Jensen, K. F. and Bawendi, M. G. 1997. (CdSe)ZnS core-shell quantum dots: Synthesis and characterization of a size series of highly luminescent nanocrystallites. *J Phys Chem B* 101:9463–75.

Dai, X., Zhang, Z., Jin, Y. et al. 2014. Solution-processed, high-performance light-emitting diodes based on quantum dots. *Nature* 515(7525):96–9.

De Merlis, C. C. and Schoneker, D. R. 2003. Review of the oral toxicity of polyvinyl alcohol (PVA). *Food Chem Toxicol* 41:319–26.

Derfus, A. M., Chan, W. C. and Bhatia, S. N. 2004. Probing the cytotoxicity of semiconductor quantum dots. *Nano Lett* 4:11–8.

Djagny, K. B., Wang, Z. and Xu, S. 2001. Gelatin: A valuable protein for food and pharmaceutical industries. *Crit Rev Food Sci Nutrition* 41:481–92

Ehlert, S., Stegelmeier, C., Pirner, D. and Forster, S. 2015. A general route to optically transparent highly filled polymer nanocomposites. *Macromolecules* 48:5323–7.

Etsuo, N., Yorihiro, Y. and Yoshiyo, K. 1978. Oxidative degradation of polymers. III. Photooxidation of poly(vinyl alcohol) in aqueous solution. In Allara, D. L., Hawkins, W. L. (Eds.), *Stabilization and Degradation of Polymers; Advances in Chemistry*. American Chemical Society: Washington, DC. 78–95. ISBN13: 9780841203815.

Flissikowski, T., Hundt, A., Lowisch, M., Rabe, M. and Henneberger, F. 2001. Photon beats from a single semiconductor quantum dot. *Phys Rev Lett* 86:3172.

Fu, Y., Kim, D., Jiang, W., Yin, W., Ahn, T. K. and Chae, H. 2017. Excellent stability of thicker shell CdSe@ZnS/ZnS quantum dots. *RSC Adv* 7:40866.

Gaponik, N., Hickey, S. G., Dorfs, D., Rogach, A. L. and Eychmüller, A. 2010. Progress in the light emission of colloidal semiconductor nanocrystals. *Small* 6:1364–78.

Gupta, J. A., Awschalom, D. D., Efros, A. L. and Rodina A. V. 2002. Spin dynamics in semiconductor nanocrystals. *Phys Rev B* 66:125307.

Gu, Z., Zou, L., Fang, Z., Zhu, W. and Zhong, X. 2008. One-pot synthesis of highly luminescent CdTe/CdS core/shell nanocrystals in aqueous phase. *Nanotechnology* 19:135604.

Ginger, D. S. and Greenham, N. C. 1999. Photoinduced electron transfer from conjugated polymers to CdSe nanocrystals. *Phys Rev B* 59:10622–9.

Grim, J. Q., Mannaab, L. and Moreels, I. 2015. A sustainable future for photonic colloidal nanocrystals. *Chem Soc Rev* 44:5897.

Harris, D. K. and Bawendi, M. G., 2012. Improved precursor chemistry for the synthesis of III–V quantum dots. *J Am Chem Soc* 134:20211–3.

Hines, M. A. and Guyot-Sionnest, P. 1996. Synthesis and characterization of strongly luminescing ZnS-capped CdSe nanocrystals. *J Phys Chem* 100:468–71.

Hong, S. P., Park, H. K., Oh, J. H., Yang, H. and Do, Y. R. 2012. Comparisons of the structural and optical properties of o-AgInS$_2$, t-AgInS$_2$, and c-AgIn$_5$S$_8$ nanocrystals and their solid-solution nanocrystals with ZnS. *J Mater Chem* 22:18939–49.

Huynh, W. U., Dittmer, J. J. and Alivisatos, A. P. 2002. Hybrid nanorod-polymer solar cells. *Science* 295:2425–7.

Iwamoto, R., Miya, M. and Mima, S. 1979. Determination of crystallinity of swollen poly(vinyl alcohol) by laser Raman spectroscopy. *J Polymer Sci: Polymer Phys Ed* 17:1507–15.

Jang, E., Jun, S., Jang, H., Lim, J., Kim, B. and Kim, Y. 2010. White light-emitting diodes with quantum dot color converters for display backlights. *Adv Mater* 22:3076–80.

Jin, G., Jiang, L. M., Yi, D. M., Sun, H. Z. and Sun, H. C. 2015. The influence of surface modification on the photoluminescence of CdTe quantum dots: Realization of bio-imaging via cost-effective polymer. *Chem Phys Chem* 16:3687–94.

Joo, S.-Y., Park, H.-S., Kim, D.-Y., Kim, B.-S., Lee, C.-G. and Kim, W.-B. 2018. An investigation into the effective surface passivation of quantum dots by a photo-assisted chemical method. *AIP Advances* 8:015017.

Kang, X., Huang, L., Yang, Y. and Pan, D. 2015. Scaling up the aqueous synthesis of visible light emitting multinary AgInS$_2$/ZnS core/shell quantum dots. *J Phys Chem* C119:7933–40.

Kang, X., Yang, Y., Wang, L., Wei, S. and Pan, D. 2015. Warm white light emitting diodes with gelatin-coated AgInS$_2$/ZnS core/shell quantum dots. *ACS Appl Mater Interfaces* 7:27713–9.

Kaul, G. and Amiji, M. 2004. Biodistribution and targeting potential of poly(ethylene glycol)-modified gelatin nanoparticles in subcutaneous murine tumor model. *J Drug Target* 12:585–91.

Kaul, G. and Amiji, M. 2005. Tumor-targeted gene delivery using poly (ethylene glycol)-modified gelatin nanoparticles: In vitro and in vivo studies. *Pharm Res* 22:951–61.

Khanna, P. K., Gokhale, R. R., Subbarao, V., Singh, N., Jun, K.-W. and Das, B. K. 2005. Synthesis and optical properties of CdS/PVA nanocomposites. *Mat Chem Phys* 94:454–9.

Kim, J.-H. and Yang, H. 2014. White lighting device from composite films embedded with hydrophilic Cu(In, Ga)S$_2$/ZnS and hydrophobic InP/ZnS quantum dots. *Nanotechnology* 25:225601.

Konstantatos, G. and Sarghent, E. S. (Eds.) 2013. *Colloidal Quantum Dot Optoelectronics and Photovoltaics*. Cambridge University Press, Cambridge, United Kingdom. ISBN: 9781139022750.

Kovalchuk, A., Huang, K., Xiang, C., Martí, A. A., and Tour, J. M. 2015. Luminescent polymer composite films containing coal-derived graphene quantum dots. *ACS Appl Mater Interfaces* 7:26063–8.

Kortan, A. R., Hull, R., Opila, R. L., Bawendi, M. G., Steigerwald, M. L., Carroll, P. J. and Brus, L. E. 1990. Nucleation and growth of CdSe on ZnS quantum crystallite. Seeds, and vice versa, in inverse Micelle media. *J Am Chem Soc* 112:1327–32.

Kozlov, P. V. and Burdygina, G. I. 1983. The structure and properties of solid gelatin and the principles of their modification. *Polymer* 24:651–66.

Kroutvar, M., Ducommun,Y., Heiss, D., Bichler, M., Schuh, D., Abstreiter, G. and Finley, J. J. 2004. Optically programmable electron spin memory using semiconductor quantum dots. *Nature* 432:81–4.

Kryshtab, T., Borkovska, L. V., Gudymenko, O., Stroyuk, O., Raevskaya, A. and Fesenko, O. 2016. Photoinduced transformations of optical properties of CdSe and Ag-In-S nanocrystals embedded in the films of polyvinyl alcohol. *AIMS Mat Sci* 3:658–68.

Kucur, E., Riegler, J., Urban, G. A. and Nann, T. 2004. Charge transfer efficiency in hybrid bulk heterojunction composites. *J Chem Phys* 121:1074–9.

Kuljanin, J., Comor, M. I., Djokovic, V. and Nedeljkovic, J. M. 2006. Synthesis and characterization of nanocomposite of polyvinyl alcohol and lead sulfide nanoparticles. *Mat Chem Phys* 95:67–71.

Kuno, M., Lee, J. K., Dabbousi, B. O., Mikulec, F. V. and Bawendi, M. G. 1997. The band edge luminescence of surface modified CdSe nanocrystallites: Probing the luminescing state. *J Chem Phys* 106:9869.

Lee, C. W., Chou, C. H., Huang, J. H., Hsu, C. S. and Nguyen, T. P. 2008. Investigations of organic light emitting diodes with CdSe(ZnS) quantum dots. *Mater Sci Eng B* 147:307–11.

Lesnyak, V., Gaponik, N. and Eychmüller, A. 2013. Colloidal semiconductor nanocrystals: The aqueous approach. *Chem Soc Rev* 42:2905–29.

Li, Y. Q., Rizzo, A., Cingolani, R. and Gigli, G. 2006. Bright white-light-emitting device from ternary nanocrystal composites. *Adv Mater* 18:2545–8.

Liang, R., Yan, D., Tian, R., Yu, X., Shi, W., Li, C., We, M., Evans, D. G. and Duan, X. 2014. Quantum dots-based flexible films and their application as the phosphor in white light-emitting diodes. *Chem Mater* 26:2595–2600.

Lifshitz, E., Dag, I., Litvin, I., Hodes, G., Gorer, S., Reisfeld, R., Zelner, M. and Minti, H. 1998. Optical properties of CdSe nanoparticle films prepared by chemical deposition and sol-gel methods. *Chem Phys Lett* 288:188–96.

Luo, Y. and Sun, X. 2007. One-step preparation of poly(vinyl alcohol)-protected Pt nanoparticles through a heat-treatment method. *Mater Lett* 61:2015–7.

Mao, B., Chuang, C.-H., Wang, J. and Burda, C. 2011. Synthesis and photophysical properties of ternary I-III-VI AgInS$_2$ nanocrystals: Intrinsic versus surface states. *J Phys Chem C* 115:8945–54.

Mansur, H. S., Mansur, A. and González, J. C. 2011a. Synthesis and characterization of CdS quantum dots with carboxylic-functionalized poly (vinyl alcohol) for bioconjugation. *Polymer* 52:1045–54.

Mansur, A., Mansur, H. and González, J. 2011b. Enzyme-polymers conjugated to quantum-dots for sensing applications. *Sensors* 11:9951–72.

Mansur, H. S., Mansur, A., González, J. C. and Chab, V. 2012. Protein-semiconductor quantum dot hybrids for biomedical applications. *Phys Stat Sol C* 9:1435–8.

Mattoussi, H., Palui, G. and Na, H. B. 2012. Luminescent quantum dots as platforms for probing in vitro and in vivo biological processes. *Adv Drug Del Rev* 64:138–66.

Meinardi, F., Colombo, A., Velizhanin, K. A., Simonutti, R., Lorenzon, M., Beverina, L., Viswanatha, R., Klimov, V. I. and Brovelli, S. 2014. Large-area luminescent solar concentrators based on 'Stokes-shift-engineered' nanocrystals in a mass-polymerized PMMA matrix. *Nature Photonics* 8:392–9.

Meulenberg, R. W., Jennings, T. and Strouse, G. F. 2004. Compressive and tensile stress in colloidal CdSe semiconductor quantum dots. *Phys Rev B* 70:235311.

Mozafari, M. and Moztarzadeh, F. 2010. Controllable synthesis, characterization and optical properties of colloidal PbS/gelatin core-shell nanocrystals. *J Colloid Interface Sci* 351:442–8.

Murata, Y., Jo, Jun-ichiro and Tabata, Y. 2017. Preparation of gelatin nanospheres incorporating quantum dots and iron oxide nanoparticles for multimodal cell imaging. *J Biomater Sci Polymer Edn* 28:555–68.

Nirmal, M., Murray, C. B. and Bawendi, M. G. 1994. Fluorescence-line narrowing in CdSe quantum dots: Surface localization of the photogenerated exciton. *Phys Rev B* 50:2293.

Nirmal, N., Norris, D. J., Kuno, M., Bawendi, M. G., Efros, A. L. and Rosen, M. 1995. Observation of the "Dark exciton" in CdSe quantum dots. *Phys Rev Lett* 75:3728.

Ozga, K., Michel, J., Nechyporuk, B. D., Ebothé, J., Kityk, I. V., Albassam, A. A., El-Naggar, A. M. and Fedorchuk, A. O. 2016. ZnS/PVA nanocomposites for nonlinear optical applications. *Physica E* 81:281–9.

Parani, S., Pandian, K. and Oluwafemi, O. S. 2018. Gelatin stabilization of quantum dots for improved stability and biocompatibility. *Int J Biol Macromol* 107 A:635–641.

Pendyala N. B. and Rao, K. S. R. K. 2009. Efficient Hg and Ag ion detection with luminescent PbS quantum dots grown in poly vinyl alcohol and capped with mercaptoethanol. *Colloid Surf A* 339:43–7.

Peng, X., Schlamp, M. C., Kadavanich, A. and Alivisatos, A. P. 1997. Epitaxial growth of highly luminescent CdSe/CdS core/shell nanocrystals with photostability and electronic accessibility. *J Am Chem Soc* 119:7019–29.

Pelley, J. L., Daar, A. S. and Saner, M. A. 2009. State of academic knowledge on toxicity and biological fate of quantum dots. *Toxicol Sci* 112:276–96.

Pezron, I., Djabourov, M., Bosio, L. and Leblond, J. 1990. X-ray-diffraction of gelatin fibers in the dry and swollen states. *J Polymer Sci Part B-Polymer Phys* 28:1823–39.

Porel, S., Singh, S., Harsha, S. S., Rao, D. N. and Radhakrishnan, T. P. 2005. Nanoparticle-embedded polymer: In situ synthesis, free-standing films with highly monodisperse silver nanoparticles and optical limiting. *Chem Mater* 17:9–12.

Pritchard, J. G. 1970. *Polyvinyl Alcohol. Basic Properties and Uses.* Macdonald & Company, ISBN: 035603335X, 9780356033358.

Raevskaya, A. E., Stroyuk, A. L. and Kuchmiy, S. Y. 2006. Preparation of colloidal CdSe and CdS/CdSe nanoparticles from sodium selenosulfate in aqueous polymers solutions. *J Colloid Interface Sci* 302:133–41.

Raevskaya, A. E., Ivanchenko, M. V., Stroyuk, O. L., Kuchmiy, S. Y. and Plyusnin, V. F. 2015. Luminescent Ag-doped In_2S_3 nanoparticles stabilized by mercaptoacetate in water and glycerol. *J Nanopart Res* 17:135.

Ramanery, F. P., Mansur, A. and Mansur, H. S. 2014. CdSe/CdS core/shell quantum dots synthesized with water soluble polymer for potential biosensor applications. *Mater Sci Forum* 805:83–8.

Ren, S., Chang, L.-Y, Lim, S.-K., Zhao, J., Smith, M., Zhao, N., Bulović, V., Bawendi, M. and Gradečak, S. 2011. Inorganic-organic hybrid solar cell: Bridging quantum dots to conjugated polymer nanowires. *Nano Lett* 11:3998–4002.

Rogach, A. L., Franzl, T., Klar, T. A., Feldmann, J., Gaponik, N., Lesnyak, V., Shavel, A., Eychmüller, A., Rakovich, Y. P. and Donegan, J. F. 2007. Aqueous synthesis of thiol-capped CdTe nanocrystals: State-of-the-art. *J Phys Chem C* 111:14628–37.

Rudko, G. Y., Kovalchuk, A. O., Fediv, V. I., Beyer, J., Chen, W. M. and Buyanova, I. A. 2012. Effects of ultraviolet light on optical properties of colloidal CdS nanoparticles embedded in polyvinyl alcohol (PVA) matrix. *Adv Sci Eng Med* 4:394–400.

Rudko, G. Y., Kovalchuk, A. O., Fediv, V. I., Ren, Q., Chen, W. M., Buyanova, I. A. and Pozina, G. 2013. Role of the host polymer matrix in light emission processes in nano-CdS/poly vinyl alcohol composite. *Thin Sol Films* 543:11–5.

Rudko, G., Kovalchuk, A., Fediv, V., Chen, W. M. and Buyanova, I. A. 2015. Enhancement of polymer endurance to UV light by incorporation of semiconductor nanoparticles. *Nanoscale Research Lett* 10:81.

Saini, I., Sharma, A., Dhiman, R., Chandak, N., Aggarwal, S. and Sharma, P. K. 2017. Functionalized SiC nanocrystals for tuning of optical, thermal, mechanical and electrical properties of polyvinyl alcohol. *Thin Sol Films* 628:176–83.

Schaller, R. D. and Klimov, V. I. 2004. High efficiency carrier multiplication in PbSe nanocrystals: Implications for solar energy conversion. *Phys Rev Lett* 92:186601.

Sui, X. M., Shao, C. L. and Liu, Y. C. 2005. White-light emission of polyvinyl alcohol/ZnO hybrid nanofibers prepared by electrospinning. *Appl Phys Lett* 87:113115.

Sun, C., Qu, R., Ji, C., Meng, Y., Wang, C., Sun, Y. and Qi, L. 2009. Preparation and property of polyvinyl alcohol-based film embedded with gold nanoparticles. *J Nanopart Res* 11:1005–10.

Sharma, D., Jha, R. and Kumar, S. 2016. Quantum dot sensitized solar cell: Recent advances and future perspectives in photoanode. *Sol Energy Mat Sol Cells* 155:294–322.

Shen, L. 2011. Biocompatible polymer/quantum dots hybrid materials: Current status and future developments. *J Funct Biomater* 2:355–72.

Shen, L.-M. and Liu, J. 2016. New development in carbon quantum dots technical applications. *Talanta* 156–157:245–56.

Skobeeva, V. M., Smyntyna, V. A., Sviridova, O. I., Struts, D. A. and Tyurin, A. V. 2008. Optical properties of cadmium sulfide nanocrystals obtained by the sol-gel method in gelatin. *J Appl Spectroscopy* 75:576–82.

Smith, M. J., Malak, S. T., Jung, J., Yoon, Y. J., Lin, C. H., Kim, S., Lee, K. M., Ma, R., White, T. J., Bunning, T. J., Lin, Z. and Tsukruk, V. V. 2017. Robust, uniform, and highly emissive quantum dot–polymer films and patterns using thiol-ene chemistry. *ACS Appl Mater Interfaces* 9:17435–48.

Su, L., Zhang, X., Zhang, Y. and Rogach, A. L. 2016. Recent progress in quantum dot based white light-emitting devices. *Top Curr Chem (Z) (Springer)* 374:42.

Talapin, D. V., Mekis, I., Gotzinger, S., Kornowski, A., Benson, O. and Weller, H. 2004. CdSe/CdS/ZnS and CdSe/ZnSe/ZnS core-shell-shell nanocrystals. J Phys Chem B, 108:18826–18831.

Todescato, F., Fortunati, I., Minotto, A., Signorini, R., Jasieniak, J. J. and Bozio, R. 2016. Engineering of semiconductor nanocrystals for light emitting applications. *Materials* 9:672.

Tomczak, N., Janczewski, D., Han, M. and Vancso, G. J. 2009. Designer polymer-quantum dot architectures. *Prog Polym Sci* 34:393–430.

Ulubayram, K., Aksu, E., Gurhan, S. I., Serbetci, K. and Hasirci, N. 2002. Cytotoxicity evaluation of gelatin sponges prepared with different cross-linking agents. *J Biomater Sci Polymer Ed* 1203–1219.

Vineeshkumar, T. V., Raj, D. R., Prasanth, S., Unnikrishnan, N. V., Philip, R. and Sudarsanakumar, C. 2014. Structural and optical studies of $Zn_{1-x}Cd_xS$ quantum dots synthesized by in situ technique in PVA matrix. *Optical Materials* 37:439–45.

Volkov, A. S., Kumekov, S. E., Syrgaliev, E. O. and Chernyshov, S. V. 1991. Photoluminescence and anti-Stokes emission of native collagen in visible range of the spectrum. *Biofizika* 36:770–3.

Walling, M. A., Novak, J. A. and Shepard, J. R. E. 2009. Quantum dots for live cell and in vivo imaging. *Int J Mol Sci* 10:441–91.

Wang, H., Chen, Z., Fang, P. and Wang, S. 2007. Synthesis, characterization and optical properties of hybridized CdS-PVA nanocomposites. *Mater Chem Phys* 106:443–6.

Wang, Y., Chen, H., Ye, C. and Hu, Y. 2008. Synthesis and characterization of CdTe quantum dots embedded gelatin nanoparticles via a two-step desolvation method. *Mater Lett* 62:3382–4.

Wang, Y., Ye, C., Wu, L. and Hu, Y. 2010. Synthesis and characterization of self-assembled CdHgTe/gelatin nanospheres as stable near infrared fluorescent probes in vivo. *J Pharmac Biomed Analysis* 53:235–42.

Wang, L., Kang, X. and Pan, D. 2016a. High color rendering index warm white light emitting diodes fabricated from $AgInS_2$/ZnS quantum dot/PVA flexible hybrid films. *Phys Chem Chem Phys* 18:31634–9.

Wang, Y., Zhao, Y., Zhang, F., Chen, L., Yang, Y. and Liu, X. 2016b. Fluorescent polyvinyl alcohol films based on nitrogen and sulfur co-doped carbon dots towards white light-emitting devices. *New J Chem* 40:8710–6.

Wang, P., Zhang, Y., Ruan, C., Su, L., Cui, H. and Yu, W. W. 2017. A few key technologies of quantum dot light-emitting diodes for display. *IEEE J Select Top Quantum Electron* 23:2000312.

Wang, X.-F., Wang, G.-G., Li, J.-B., Liu, Z., Zhao, W.-F. and Han, J.-C. 2018. Towards high-powered remote WLED based on flexible white-luminescent polymer composite films containing S, N co-doped graphene quantum dots. *Chem Eng J* 336:406–15.

Xu, L., Chen, K., El-Khair, H. M., Li, M. and Huang, X. 2001. Enhancement of band-edge luminescence and photo-stability in colloidal CdSe quantum dots by various surface passivation technologies. *Appl Surf Sci* 172:84–8.

Yakimets, I., Wellner, N., Smith, A. C., Wilson, R. H., Farhat, I. and Mitchell, J. 2005. Mechanical properties with respect to water content of gelatin films in glassy state. *Polymer* 46:12577.

Yannas, I. V. 1967. Cross-linking of gelatine by dehydration. *Nature* 215:509–10.

Zhang, S., Geryak, R., Geldmeier, J., Kim, S. and Tsukruk, V. V. 2017. Synthesis, assembly, and applications of hybrid nanostructures for biosensing. *Chem Rev* 117:12942–3038.

Zhong, H., Bai, Z. and Zou, B. 2012. Tuning the luminescence properties of colloidal I–III–VI semiconductor nanocrystals for optoelectronics and biotechnology applications. *J Phys Chem Lett* 3:3167–75.

CHAPTER 4

Development of Multifunctional Nanocomposites by Cavitation

Rada Savkina

CONTENTS

4.1 BASIC CONCEPTS

The effect of ultrasound (US) based on the acoustic cavitation phenomenon is found to be a new, environmentally friendly strategy for the functional materials fabrication.

4.1.1 What Is a Cavitation Phenomenon?

The cavitation phenomenon is a result of local pressure reduction producing the liquid discontinuity effect, which occurs either with an increase in the liquid velocity (*hydrodynamic cavitation*) or during propagation of an acoustic wave into the half-period of rarefaction (*acoustic cavitation*), when the acoustic power is enough to overcome the molecular bonding forces in the liquid. The acoustic cavitation in the liquid has a threshold character and, as illustrated in Figure 4.1, consists virtually in generation and growth of bubbles, followed by implosive collapse. Cavitation bubbles oscillate with the applied sound field and grow through a slow pumping of gas/vapour from the bulk liquid into their interior (*rectified diffusion*). The growing cavity eventually reaches a critical size where it can efficiently absorb US energy. This critical size depends on the liquid properties and the frequency of sound.

The completion phase of the cavitation oscillating – implosive collapse – is very localised and transient. During collapse, *hot spots* inside collapsing bubbles generate a temperature above 5000 K and pressures exceeding 1000 atmospheres, with heating and cooling rates in excess of 10^{10} K s^{-1} (Suslick et al. 1999). The "hot spot" approach is the most widely applied to explain the chemical effect of ultrasounds. However, the supercritical phase existence in an ultrasonically irradiated solution (Hua 1995) as well as an electrical (Margulis 1992) and plasma discharge (Nikitenko and Pflieger 2017) theory of cavitation is used to justify the sonochemical processes in cavitating fluid. In any case, there is a general consensus that the chemical and physical effects of power ultrasound are related to extremely rapid implosion of the cavitation bubbles occurring at the final stage of collapse.

4.1.2 System Parameters and Effectiveness of the Cavitation

It should be noted that acoustic frequency and power, as well as the liquid nature, liquid height and liquid temperature applied pressure, dramatically affect the chemical and physical effects of power ultrasound. The geometry of the reactor and a source of high-energy vibrations are essential components for the initiation of the cavitation phenomenon also.

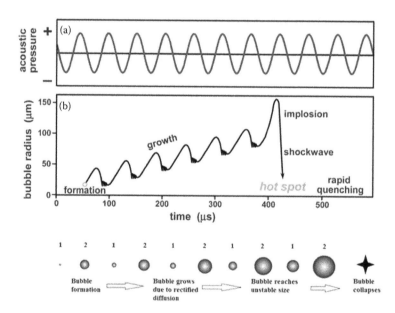

FIGURE 4.1 Schematic illustration of the process of acoustic cavitation: the formation, growth and implosive collapse of bubbles in a liquid irradiated with high intensity ultrasound: 1 – the half-period of rarefaction; 2 – the half-period of compression. (Adapted with permission from Xu et al., 2013. Copyright 2012, The Royal Society of Chemistry.)

It should be noted the latest well-grounded review analysed sonochemical activity in aqueous solutions through listed system parameters (Wood et al. 2017). Each of the parameters are considered in terms of their effect on cavitation bubble characteristics and the subsequent impact on sonochemical activity. A sonochemical system can be configured through bubble structural instability. This would allow one to increase or control sonochemical reaction effectiveness.

Cavitation occurs over a very wide range of US frequencies – from tens of Hz to MHz. The frequency area involving low-frequency (high-power) acoustic waves, known as "power ultrasound", lies between 20 and 100 kHz, and it is used for sonochemistry and for such fields as US cleaning and US welding. The frequency range available for sonochemistry has been extended to 2 MHz with the development of high power equipment capable of generating cavitation (Mason and Lorimer 2002). US frequency above 2 MHz does not produce cavitation at ordinary conditions in consequence of the intrinsic viscosity of liquids. This region, with high frequency and low power, is used for analytical purposes and in medical imaging.

At the same time, there is a way to increase the acoustic power in the high frequency region. *Focusing of acoustic pulses* is a fundamental aspect of most of the medical and industrial applications of high-frequency ultrasound. Focused ultrasound uses an acoustic lens to concentrate multiple intersecting US beams on a target where the US energy can have useful thermal or mechanical effects. One of its applications is known as High Intensity Focused Ultrasound (HIFU) (Dubinsky et al. 2008).

There are three main types of a source of high-energy vibrations: the liquid whistle, and two electromechanical devices based on the magnetostrictive and the piezoelectric effects (Mason 1999). A fourth type of transducer, the magnetically driven vibrating bar, generates very high-power vibrations but in the audible range. Thus, most investigators use a high-intensity US horn, operating at a frequency between 20 and 55 kHz, for sonochemical experiments (Table 4.1). A typical laboratory sonochemical reactor is presented in Figure 4.2.

It was found that for the cavity, considering isothermal growth followed by adiabatic collapse, energy E_c ($E_c = W/t_c$, where W is the total work done by the surroundings on the cavity during the isothermal expansion and the adiabatic compression phases, t_c is the lifetime of the cavity) released by the bubble at the end of collapse increases with an increase in the frequency and decreases with an increase in the intensity for an individual cavity, containing a vapour-saturated gas mixture, as well as for a gaseous cavity (Vichare et al. 2000). At the same time, the threshold intensity for cavitation, defined as the acoustic pressure amplitude, at which a cavitation event is first detected, is increased with frequency rise (see, for example, Sponer 1990). At a very high frequency (>2 MHz), the time required for the rarefaction cycle is too short to permit a cavitation bubble to grow to a size sufficient to cause optimal disruption of the liquid.

Observations reported in the literature indicate that the cavities at lower frequencies are inefficient in utilising the energy of the ultrasound. Experimental results using pulsed ultrasound (Guiterez and Henglein 1990) show that iodine liberation drops to zero with an increase in intensity. Sonochemical degradation of aqueous carbon tetrachloride at various frequencies indicated higher efficiencies at 500 kHz than at 20 kHz (Francony and Petrier

TABLE 4.1 Manufacturers and Parameters of US Equipment for Liquid Processing

Manufacturer	US Equipment for Liquid Processing	US Converter and Frequency	Output Power/Acoustic Intensity
Sonics & Materials Inc. www.sonics.com	Vibra-cell processors (VCX 130, VCX 500, VCX 750, VCX 2500)	Titanium horn (Ti-6Al-4 V) 20 kHz	130–2500 W/ 32 Wcm^{-2} for 750 W
Hielscher Ultrasonics GmbH www.hielscher.com	Powerful and Versatile Homogeniser UIP1000hdT	Titanium horn BS2d22 20 kHz	1000 W/57 Wcm^{-2}
Branson Ultrasonics www.emerson.com	SFX150 Sonifier SFX250 Sonifier SFX550 Sonifier	Titanium horn 40 kHz 20 kHz 20 kHz	150 W 250 W 550 W
Hangzhou Dowell Ultrasonic Technology Co. www. dowellsonic.com	DW-SL40-100 DW-SL40-200 DW-SL28-300 DW-SL28-800 DW-SL20-1000	Titanium horn 40 kHz/55 kHz 28 kHz/40 kHz 28 kHz/40 kHz 28 kHz 20 kHz	100–1200 W/5 MPa

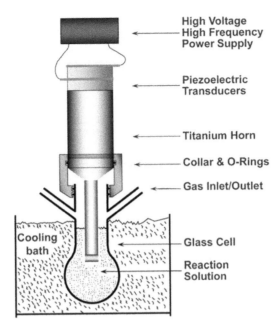

High Voltage
High Frequency
Power Supply

Piezoelectric
Transducers

Titanium Horn

Collar & O-Rings

Gas Inlet/Outlet

Cooling
bath

Glass Cell

Reaction
Solution

FIGURE 4.2   A typical laboratory rig for sonochemical reactions uses a high-intensity ultrasonic horn, typically around 20 kHz and 10 to 100 W acoustic power. (Reprinted with permission from Xu et al., 2013. Copyright 2012, The Royal Society of Chemistry).

1996). Whereas the physical effects of an ultrasound (for example, surface damage) are more dominant at lower frequencies, and cavitational heating of collapsing bubbles occurs over the full frequency range.

Thus, for the overall success of sonochemical processes, a lower intensity and higher frequency of the ultrasound is always favourable (see Figure 4.3). But, a bubble contains not only the gas that is dissolved in the liquid, but also vapour from the liquid itself. The amount of vapour in the bubble depends on the vapour pressure of the liquid, which is strongly dependent on the temperature of the bulk liquid. Higher temperature and lower applied pressure lead to an increase of the vapour pressure of the solvent; thereby, the efficiency of the collapse of the cavity decreases.

4.2 ULTRASOUND-ASSISTED METHODS FOR MATERIALS FABRICATION

The application of ultrasound for a new functional materials fabrication was discussed by early remarkable reviews (Gedanken 2004; Bang and Suslick 2010; Xu et al. 2013; Hinman and Suslick 2017) and books (Manickam and Ashokkumar 2014; Chatel 2017). Scientific interest in this field has considerably increased and focused on various topics such as preparation of nanoparticles, porous and nanostructured materials, nucleation processes, semiconductor nanostructure synthesis, etc.

We can demonstrate a striking example of ultrasound resources in the area of nano-material fabrication from Zak et al. (2013). Figure 4.4 shows SEM images and schema of a fast and easily realisable, as well as "green", synthesis of various shaped ZnO nanocrystals. Ultrasound-assisted synthesis produces high-quality hierarchical ZnO

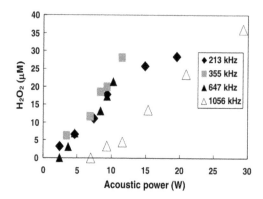

FIGURE 4.3 Variation in H_2O_2 yield in water with acoustic frequency and power. It is generally accepted that the yield of H_2O_2 can be considered an indicator of sonochemical activity. (Reprinted with permission from *Ultrason Sonochem*, 15, Kanthale et al., Sonoluminescence, sonochemistry (H_2O_2 yield) and bubble dynamics: Frequency and power effects, 143–50. Copyright 2008, Elsevier.)

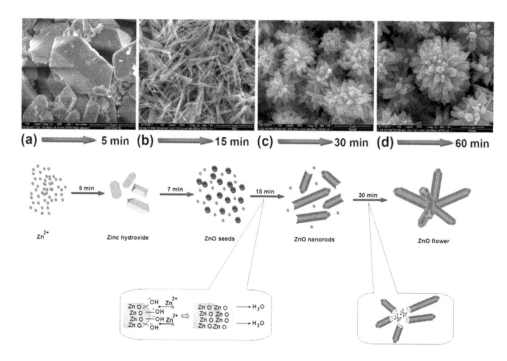

FIGURE 4.4 SEM micrographs (a–d) and illustrating diagram (e) for the morphological evolution from the newly formed (a) $Zn(OH)_2$ crystals at formation time of 5 min ultrasonocation, (b) ZnO nanorods and (c, d) ZnO flowers prepared at the following times of 15, 30, and 60 min ultrasonocation. (Adapted with permission from *Ultrason Sonochem*, 20, Zak et al., Sonochemical synthesis of hierarchical ZnO nanostructures, 395–400, Copyright 2013, Elsevier.)

nanostructures with controllable shapes, uniformity and purity from Zn salt, sodium hydroxide and ammonia solution without other structure-directing agent or surfactant, high temperatures and/or highly toxic chemicals. Note that ZnO-based materials are used in many fields because of their photocatalytic nature, low cost and environmental sustainability.

4.2.1 Sonochemical Synthesis and Ultrasonic Spray Pyrolysis

Ultrasound-assisted methods of new materials fabrication, such as *sonochemical synthesis* (SCS) and *ultrasonic spray pyrolysis* (USP), are based on chemical and physical effects following cavitation bubble collapse. Chemical effects are driven from the hot spot phenomenon inside collapsing bubbles. The physical effects are caused by ultrasound-induced shock waves, microjets at the liquid–solid interface and secondary phenomena such as turbulence, rapid mixing and interparticle collisions in slurries (Hinman and Suslick 2017).

Sonochemical synthesis is a direct method based on the primary sonochemistry reactions occurring inside bubble and secondary chemical US effects, where sonochemical products are formed inside the bubble but then diffuse into the liquid phase and subsequently react with solution precursors to form nanostructured materials (Bang and Suslick 2010; Xu et al. 2013).

Most research in this field has focused on II-VI semiconductors – nanostructured chalcogenides such as CdS, ZnS, PbS, CdSe, ZnSe, PbSe, etc. The reason for their popularity is in their importance to nonlinear optics, photovoltaic fields and optoelectronics, and for such technological applications as biological labels and electrochemical cells. A typical sonochemical synthesis of II-VI semiconductors involves the US irradiation of an aqueous solution of a metal salt and a chalcogenide source (Bang and Suslick 2010).

The advantages of the sonochemical approach over conventional methods in the synthesis of metallic and alloyed nanoparticles (NPs), such as Au, Co, Fe, Ni, Ag, Pd, Au/Pd, Fe/Co (see Gedanken 2004; Shchukin 2010), and metal oxides, such as TiO_2, ZnO, CeO_2, MoO_3, V_2O_5, In_2O_3, $ZnFe_2O_4$, $PbWO_4$, $BiPO_4$ and $ZnAl_2O_4$ (see Bang and Suslick 2010), have been recognised by many research groups. It should be noted that the sonochemical synthesis of high surface area carbon materials was reviewed recently by Skrabalak (2009).

If sonochemical synthesis generally uses high-intensity ultrasound with a low frequency (kHz), such methods as US spray pyrolysis (e.g. Hinman and Suslick 2017), based on the physical effects of ultrasound, utilise MHz frequency range (near 2 MHz) and consist in nebulisation precursor solutions to produce the micron-sized droplets that act as individual micron-sized chemical reactors. Spray pyrolysis has been widely used in industry for film deposition as well as micron-sized powders of metals, metal alloys and ceramic materials, metal oxides and metal chalcogenides.

4.2.2 Sonocrystallisation and Sonofragmentation

One should also note such promising techniques, especially for the pharmaceutical industry, as *sonocrystallisation* and *sonofragmentation* (Sander et al. 2014), as well as the effects of ultrasonically *enhanced polymerisation* (Zhang et al. 2009).

4.2.3 Coupling Cavitation with Other Techniques

New possibilities are opened by coupling cavitation with other techniques, such as microwave (Martina et al. 2016) or microfluidics (Rivas and Kuhn 2016), that facilitate the design of substantially cleaner, safer and more energy-efficient materials science and technologies. Synergetic action of ultrasound and light radiation or electrical fields results in new powerful methods of material fabrication, which include *sonophotodeposition* and *sonoelectrodeposition* processes. Examples of different materials prepared by sonoelectrodeposition and sonophotodeposition methods are presented in Magdziarz and Colmenares' (2017) review. Among them are Pt nanoflowers and fractal structures for non-enzymatic sensors of hydrogen peroxide and glucose; metallic and alloyed NPs (e.g. In, Co-Pt, FePt, FePd) with potential applications in microelectronics, optical, electronics and magnetic devices, as well as industrial application; and Ag and Cu-Ag NPs with bactericidal properties against *Staphylococus aureus* and *Escherichia coli* bacteria.

At the same time, the possibilities of ultrasound-assisted methods are not limited to the fabrication of ordinary nanostructures. This capability is the basis for creating more complex systems, which are composite and hybrid materials. This problem will be described in the next section.

4.3 SYNTHESIS OF NANOCOMPOSITES AND HYBRID STRUCTURES BY CAVITATION

4.3.1 Several Topics Related to the Complex Structures Fabrication

Early reviews (Gedanken 2004; Bang and Suslick 2010; Skorb and Möhwald 2016) describe several topics related to complex structures fabrication. These topics are:

- *deposition of nanoparticles onto different surfaces, including mesoporous materials;*

- *core-shell structure fabrication; and*

- *polymer-based hybrid nanocompounds formation.*

The deposition of NPs onto the surface of substrates, as well as the insertion of nanomaterials into mesoporous materials, is directly conditioned by the action of the ultrasound-induced shock waves and microjets at the liquid–solid interface. For example, NPs (e.g. amorphous nanosized catalysts) are deposited as a smooth layer on the inner mesopores' walls without blocking them.

Another advantage of the US-assisted synthesis of nanocomposites is obtaining various forms of mesostructured silica spheres, e.g. core-shell structures SiO_2/MoS_2 and SiO_2/TiO_2 (Suh et al. 2006). On full etching, an inner silica core is gone, and only a porous TiO_2 spherical shell remains. The microspheres loaded with a potential drug for Alzheimer's disease showed an outstanding ability to selectively deliver the drug to cytosol rather than cell nuclei. The porous MoS_2 is shown to be an extremely active catalyst for hydrodesulphurisation reactions compared to its nonporous counterpart. Sonochemical deposition of noble metal NPs on substrates such as SiO_2 or metal sulphides, should also be noted. For example, sonochemically synthesised core-shell $Fe_3O_4/SiO_2/Ag$ nanocubes showed a high efficiency in the catalytic reduction reaction of 4-nitroaniline to 4-phenylenediamine and a better performance than both Ag and SiO_2/Ag nanoparticles (Abbas et al. 2015).

Oxide-based nanocomposites can be simply sonochemically produced, not only in core-shell configuration. Figure 4.5 shows highly photocatalytically active nanocomposites SiO_2/Ho_2O_3 (Zinatloo-Ajabshir et al. 2017). The role of the US time on grain size and shape of SiO_2/Ho_2O_3 structure has been demonstrated. The ideal time for production of uniform, sphere-like SiO_2/Ho_2O_3 nanocomposites with a finer size is 10 min. By altering the sonication time

FIGURE 4.5   FESEM images of SiO_2/Ho_2O_3 nanocomposite obtained with the aid of 0.5 mole of Ho source and Tetramethylethylenediamine as a basic agent at 10 (a and b) and 15 (c and d) min, and (e) without sonication. (Reproduced with permission from *Ultrason Sonochem*, 39, Zinatloo-Ajabshir et al., Simple sonochemical synthesis of Ho_2O_3-SiO_2 nanocomposites as an effective photocatalyst for degradation and removal of organic contaminant, 452–60, Copyright 2017, Elsevier.)

from 10 to 15 min, less uniform nanobundles with larger grain sizes are produced, owing to the Ostwald ripening process. Also, SiO_2/Ho_2O_3 samples obtained via vigorous stirring without ultrasound illustrate bulk structures/high agglomerated microstructures (Figure 4.5e).

The next approach for US-assisted complex structure engineering is hybrid polymer matrix composites fabrication. In the area of polymer science, the extreme conditions produced by ultrasound are known to act as a special initiator to allow chemical bonds to break and thus enhance polymerisation. Specific examples of hybrid nanocompounds, such as ZnO/PBMA,* PBMA/ZnO/PANI and Ag/PANI synthesised via ultrasound, assisted *in situ* emulsion polymerisation processes, and anticorrosion properties of such formulated coatings were described by Sonawane et al. (2014). Meanwhile, long-continued action of ultrasound leads to the degradation of polymer chains, resulting in a low molecular weight at the end (Poinot et al. 2013). At the same time, US influence has been effectively adopted for rupturing high molecular weight polymer chains, especially different polysaccharides like xanthan gum and guar gum.

Polymer-matrix composites containing inorganic fillers (filler/polymer composites) have been receiving significant attention lately because of their interesting and useful characteristics, such as good mechanical properties, thermal resistance and chemical reagent resistance. Control of nanoscale dispersion of filler is one of the important points for the fabrication of nanocomposites that can be achieved by ultrasonication and modification techniques. Moreover, compared with conventional polymerisation, it can be regarded that sonochemical activation offers some attractive features, such as low reaction temperatures, faster polymerisation rates and higher molecular weight of polymers.

The sonochemical approach to fabrication of polymer-based hybrid with carbon nanotubes (CNTs) (Zhang et al. 2009) and metallic NPs (Skorb and Möhwald 2016) has shown multiple advances. Embedded in a polymer matrix, CNT can be used as strain sensors at a nanoscale level. CNTs/PMMA composite thin films are used for gas-sensing applications. Ultrasound-assisted fabrication of the core-shell typed microbeads with CNT coatings allows them to obtain complex conducting hybrid-type nanocomposites (carboxylic acid functionalised multiwalled CNT adsorbed onto PMMA microspheres (Zhang et al. 2009)) or magnetic nanocomposite structures (Carbonyl iron/CNT (Fang and Choi 2008)). The polymer-metal hybrid nanocompounds can be applicable for biomedicine as a component of self-regulation and biomimetic systems, as well as in the automotive or aircraft industry for noise or vibration-damping components. Conducting polymers such as PPy or polyaniline could passivate the metal substrates and therefore prevent their degradation in aggressive media and produce interfaces with self-regulating properties. For example, the PPy/aluminium system constructed via ultrasonication (Skorb et al. 2013) shows high potential as a component of self-regulation with self-healing and switchable hydrophilicity properties.

4.3.2 Recent Achievements

The goal of further discussion is to scan the more recent (2017–2018) achievements for solid state nanocomposites and hybrid systems synthesised using US radiation that was not presented in the previously mentioned reviews. Table 4.2 demonstrates that the main tendencies

---

* Note that the abbreviations of polymers mentioned further decrypted in Table I (Appendix).

TABLE 4.2 The Main Tendencies in the Field of Creating Multifunctional Composite and Hybrid Structures by Sonochemical Synthesis and Related US Methods

Type of Material	Composite/Hybride Structures	Method	Application/Test Reaction	References
Simple metal oxides (TiO_2, CeO_2, ZrO_2, MnO_2, Mn_3O_4, Cu_2O, etc.)	CeO_2/NiO, CeO_2/ZnO	SCS	Photocatalytic activity	Farghali et al. (2018); Shah et al. (2017)
	Nd_2O_3/SiO_2	SCS	Decomposition of methyl violet contaminant under UV illumination	Zinatloo-Ajabshir et al. (2018)
	Ho_2O_3–SiO_2	SCS	Photocatalytic activity, destruction of methylene blue pollutant under UV illumination	Zinatloo-Ajabshir et al. (2017)
	TiO_2/NiO	US-assisted wet impregnation method	Photocatalytic activity	Vinoth et al. (2017)
	$CuO/Cu_2O/Cu$	SC combined thermal synthesis	Photocatalytic activity	Mosleh et al. (2018)
	ZnO/ZrO_2	Sol-gel approach under US irradiation	Pseudocapacitive material for energy storage applications	Aghabeygi et al. (2018)
	Ba/α-Bi_2O_3/γ-Fe_2O_3	US co-precipitation	Photocatalytic activity under solar light	Ramandi et al. (2017)
	Petal array-like Cu/Ni oxide composite foam	SCS	A pseudocapacitive material for energy storage	Karthik et al. (2017)
	Ni-Co/Al_2O_3-ZrO_2	US-assisted impregnation	Catalytic activity in dry reforming of CH_4	Mahboob et al. (2017) Shamskar et al. (2017) Jo and Yoo (2018)
	NiO-Al_2O_3	US co-precipitation	Employed in dry reforming of CH_4	
	g-C_3N_4/Ag/black TiO_2	SCS	Photocatalytic pollutant degradation and hydrogen production	
	Au-Fe_3O_4 NP loaded on AC	SCS	Application in water treatment	Bagheri et al. (2018)
Mixed oxides ($LaNiO_3$, $NiTiO_3$, $BaZr_yTi_{1-y}O_3$, $CoTiO_3$, etc.)	Cd_2SiO_4/Graphene	SCS	High electrochemical hydrogen storage capacities	Masjedi-Arani and Salavati-Niasari (2018)
	$LiFePO_4/C$	US-intensified micro-impinging jetting reactor	Cathode material	Dong et al. (2017)
	Ag_3PO_4/rectorite	US co-precipitation	Photocatalyst under visible light	Guo et al. (2017)
	$YbVO_4/CuWO_4$	SCS	Photocatalyst under visible light	Eghbali-Arani et al. (2018)
	$FeVO_4/V_2O_5$	SCS	Photocatalyst under UV and visible light	Ghiyasiyan-Arani et al. (2017)
	$Bi_2Sn_2O_7$–C_3N_4	US-assisted dispersion method	Nanophotocatalyst	Heidari et al. (2018)

(Continued)

TABLE 4.2 (CONTINUED) The Main Tendencies in the Field of Creating Multifunctional Composite and Hybrid Structures by Sonochemical Synthesis and Related US Methods

Type of Material	Composite/Hybride Structures	Method	Application/Test Reaction	References
Graphene-, GO- and rGO-based composites	Ag NPs decorated Graphene sheets	SCS	Catalytic performance and antibacterial application	Ganguly et al. (2017)
	PVAC/GO	Ultrasonic-microwave synergistic effects	Determination of mercuric ions from foods and environmental waters samples	Song et al. (2018)
	Fe_3O_4@GO/2-PTSC	US-assisted solid phase extraction	Excellent catalytic performance for the production of liquid fuel in Fischer-tropsch synthesis	Keramat and Zare-Dorabei (2017)
	rGO/Fe_3O_4	SCS	The production of liquid fuel	Abbas et al. (2018)
	Yb_2O_3/rGO	US bath	High-performance energy applications	Naderi et al. (2016)
	ZnO hollow microspheres/rGO	SCS	Enhanced Sunlight Photocatalytic Degradation of Organic Pollutants	Hanan H. Mohamed (2017)
	Sm_2O_3/rGO	SCS	High-performance supercapacitor	Dezfuli et al. (2017)
	Graphene-Ce-TiO_2 and Graphene-Fe-TiO_2	SCS	Application in degradation of crystal violet dye	Shende et al. (2018)
CNT-based composite and hybride structures	PAA-PVI/MWCNTs	SCS	Biosensor for measuring glucose	Jeon et al. (2017)
Polymer-based hybride structures	P(AAm-co-IA)/MWCNTs	SCS	Superabsorbent hydrogel: Swelling behaviour and Pb (II) adsorption capacity	Mohammadi Nezhad et al. (2017)
	TiO_2/MWCNT	SCS	Exhibits superior anticorrosion and mechanical performance	Kumar et al. (2018)
	epoxy-CNT	US stirring	As the structural and functional materials in humid environments	Goyat et al. (2017)
	Au, Ag and Pd NPs on functionalised MWCNT	SCS	For removal of organic dye	Moghaddari et al. (2018)

(Continued)

TABLE 4.2 (CONTINUED) The Main Tendencies in the Field of Creating Multifunctional Composite and Hybrid Structures by Sonochemical Synthesis and Related US Methods

Type of Material	Composite/Hybride Structures	Method	Application/Test Reaction	References
	rec-PET/MWCNT-BSA		To remove Pb^{2+} from water	Mallakpour and Behranvand (2017)
	PMMA/Fe_3O_4	Sonochemical oxidation and emulsion polymerisation	Enhanced thermal, mechanical, electrical and magnetic properties of the nanocomposites	Poddar et al.(2018)
	PVA-PVP/CuO-VB_1	SCS	Antibacterial activity	Mallakpour and Mansourzadeh (2018)
	Fe_3O_4-guargum	SCS	Catalytic reduction of p-nitroaniline	Balachandramohan et al. (2017)
	PET/Fe_3O_4NPs modified by CA & AS	SCS	Wastewater treatment, catalysts, biomedicine and drug delivery system and electromagnetic devises	Mallakpour and Javadpour (2018)
	PAI/val-MWCNT	US-assisted homogenisation	Biodegradable and bioactive properties	Mallakpour et al. (2017b)
	PVA/PVP/α-MnO_2-stearic acid	US-assisted stirring	Adsorbing Cd^{II} ion	Mallakpour and Motirasoul (2018)
	PVC/Tm-MWCNT	US-assisted stirring	Sensors and gas storage	Mallakpour et al. (2017a)
	Glycerol plasticised-starch (GPS)/ ascorbic acid (AA)-MWCNTs	US-assisted stirring	Removing methyleneblue from the water	Mallakpour and Rashidimoghadam (2018)
MOF	Amide-functionalised MOF	SCS	Sensing of nitrophenol, nitroaniline and nitrobenzene in acetonitrile solution	Gharib et al. (2018)
	Ni-MOF/GO	US-assisted ball milling	Adsorption of Congo red	Zhao et al. (2017)
	Amide-functionalised GO-MOF	SCS	The adsorption of methylene blue from aqueous solution	Tanhaei et al. (2018)

in the field of creating multifunctional composite and hybrid structures by SCS and related US methods are the same as those observed recently and mentioned in Section 4.3.1.

First of all, it should be noted that researchers continue to look at ways to improve the process for the *sonochemical synthesis of nanostructured simple and mixed metal oxides*. Among these are highly active mesoporous TiO_2, CeO_2, thermally stable ZrO_2, mesoporous MnO_2 and Mn_3O_4 nanoparticles, widely used as catalysts or energy storage materials; Cu_2O nanoparticles, used as reducing and structure-directing agents; and mixed metal oxide compounds with general formula ABX_3 ($LaNiO_3$ perovskite exhibited a higher catalytic activity and nanoporous $NiTiO_3$ rods, nanocrystals of $BaZr_yTi_{1-y}O_3$, $CoTiO_3$ nanocrystals for Li-ion batteries, magnetic recorders and catalysts). The sonochemistry method enables the synthesis of oxides-based nanocomposites with core-shell morphology that are often unachievable by traditional methods or surpassed in the known materials with shape, size and nano/microstructure control under fast reaction conditions as well as superior photocatalytic activity (Valange et al. 2018).

The violent development of graphene-based technology as well as preparation methods for graphene- (G-) like materials, such as highly reduced graphene oxide (hrGO) via reduction of graphite oxide, offers a wide range of possibilities for the preparation of *carbon-based inorganic nanocomposites* by the incorporation of various functional nanomaterials for a variety of applications.

Since its discovery, graphene has met with significant attention due to its unique electronic, mechanical and thermal properties, such as a charge-carrier mobility of 250,000 cm^2 $V^{-1}s^{-1}$ at room temperature, a thermal conductivity of 5000 W m^{-1} K^{-1}, an electrical conductivity of up to 6000 S cm^{-1} and a large theoretical specific surface area of 2630 m^2 g^{-1} (Khan et al. 2015). In addition, graphene is highly transparent towards visible light and the strongest material ever measured. Moreover, it could be considered a key building block of all other carbon allotropes (e.g. fullerenes and carbon nanotubes, Figure 4.6).

The architectures of graphene-based composites can be classified in the following way (Khan et al. 2015):

- graphene sheets form a continuous phase and act as a substrate for supporting a second component, which is typically an inorganic nanoparticle (e.g. metals, metal oxide, or CNTs), as well as polymeric nanostructures;

- graphene sheets act as nanofillers, incorporated into the continuous matrix of the second component.

Examples of such nanocomposites obtained by the US-assisted method are presented in Table 4.2. We can see that the most common applications of graphene- and hrGO-based nanocomposites are energy storage, sensing and catalysis. Considering its superior electron mobility and highly specific surface area, graphene can be expected to improve the photocatalytic performance of semiconductor photocatalysts such as metal oxides (e.g. TiO_2, ZnO, Cu_2O, Fe_2O_3, NiO, WO_3), metal sulphides (e.g. ZnS, CdS, MoS_2), metallates (e.g. Bi_2WO_6, $Sr_2Ta_2O_7$, $BiVO_4$, $InNbO_4$ and $g-Bi_2MoO_6$) and other nanomaterials (e.g. CdSe, Ag/AgCl, C_3N_4)

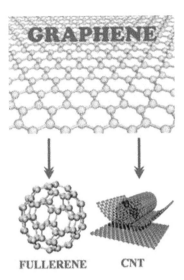

FIGURE 4.6 Graphene as a two-dimensional building block for carbonaceous materials of all other dimensions. (Adapted from Khan et al., 2015 with permission. Copyright 2015, The Royal Society of Chemistry.)

where graphene act as an efficient electron acceptor to enhance the photo-induced charge transfer and to inhibit the recombination of the nonequilibrium charge carriers. The exploitation of graphene and graphene-based materials for the fabrication of electrochemical sensors and biosensors has gained importance. Graphene-based inorganic nanocomposites have also been applied for waste water treatment or gas sorption.

Sonochemistry is useful in the synthesis of other carbon-based nanocomposite and hybrid materials as well. Among them are multiwalled carbon nanotubes (MWCNTs) and modified carbon nanotubes, with their excellent electrical conductivity, high surface area, good corrosion resistance and unique chemical, physical and mechanical properties, which make them unique and one-dimensional macromolecules – building blocks of nanotechnology for electrochemical biosensors fabrication (Ahammad et al. 2009; Zhu et al. 2012). Furthermore, due to their high reactivity and large micropore volume, MWCNTs are known as efficient, eco-friendly adsorbents for the removal of organic dyes and heavy metals from water (e.g. Figure 4.7a shows SEM image of the MWCNT-based nanocomposite which demonstrates promising characteristics for removal of organic dye). Sonochemical technology has attracted growing attention in the combination of organic and inorganic materials because it can provide conditions to achieve a very high homogeneity dispersed MWCNTs (and individual CNTs) in solution, a clean media and fine dispersion of fillers in a short time.

Applications of ultrasound to solid state composite and hybrid systems fabrication have been continually extended to polymers-based structures (see Table 4.2). The principal factors that govern enhancement in the physical properties of polymer matrix nanocomposites are the extent of encapsulation and nature of the dispersion of nanofillers in a polymer matrix. High-intensity microconvection/micro-turbulence generated by the physical effect of ultrasound during SCS of the polymer-based nanocomposites causes a very uniform

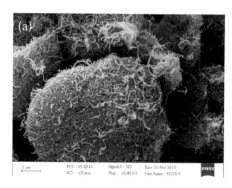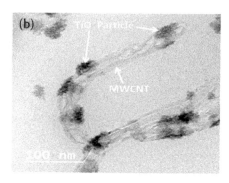

FIGURE 4.7 (a) FESEM images of the MWCNT/Ag nanocomposite and (b) TEM images of the MWCNT/TiO$_2$ nanocomposite. (Adapted with permission from *Ultrason Sonochem*, 42, Moghaddari et al., A simple approach for the sonochemical loading of Au, Ag and Pd nanoparticle on functionalized MWCNT and subsequent dispersion studies for removal of organic dyes: Artificial neural network and response surface methodology studies, 422–33 and from *Ultrason Sonochem*, 41, Kumar et al., MWCNT/TiO$_2$ hybrid nano filler towards high-performance epoxy composite, 37–46, Copyright 2018, Elsevier.)

mixing and significant improvement in the dispersion of nanofillers in the polymer matrix as well as prevents the agglomeration of the polymer nanocomposite latex particles. For example, the use of sonication during the synthesis of PMMA/Fe$_3$O$_4$ nanocomposites (Poddar et al. 2018) has resulted in a significantly smaller mean size and a narrower size distribution of Fe$_3$O$_4$ particles, as compared to particles synthesised with mechanical stirring. Moreover, authors demonstrate uniform encapsulation and dispersion of the Fe$_3$O$_4$ NPs in a PMMA matrix at relatively low loading of $\leq$5 wt % due to intense micro-turbulence generated in the reaction medium.

Hybrid polymer composites reinforced with CNTs have recently received great interest because they often contain improved properties over polymers (e.g. Lee et al. 2011). CNT/polymer composites exhibit a great strengthening effect. However, agglomeration or poor affinity of CNT in composite has limited their application. One of the commonly used approaches for uniform dispersion of CNTs in polymer nanocomposites is covalent functionalisation. The sonochemical technique used here is an accessible and efficient technique with good results (e.g. see papers of Mallakpour et al. 2017a and Mallakpour et al. 2017b).

It should also be noted that sonochemical deposition of nanoparticles onto mesoporous materials, discussed in the previous section, moves in the direction of porous coordination polymers (PCPs), which are popularly known as metal-organic frameworks (MOFs). Over the last decade, progress in the development of MOFs-based structures for gas storage, separation, catalysis, energy, sensing and biomedical applications has moved in the PCPs direction, drawing a great deal of attention due to their highly functional properties, such as their crystalline nature, high chemical and thermal stability, high surface area and excellent optical properties (Kumar et al. 2017). The MOFs have shown a significant role in the development of new water-applicable adsorbents.

The structure of metal-organic frameworks is made up of metal-oxide clusters linked by organic linkers via strong covalent bonds. These materials have high pore volume, highly ordered pore structure, high density of active sites and high specific surface area, which provide many advantages to their applications in water treatment technologies. Among many different approaches (e.g. hydro/solvothermal, microwave, electrochemical, mechanochemical, layer-by-layer and high-throughput syntheses) using aqueous and non-aqueous media for the synthesis of MOFs, the sonochemical synthesis and sonocrystallisation have been employed, providing a number of advantages (e.g. more facile preparation, fast kinetics, high phase purity, high yield, low cost and commercially viable routes).

Thus, there are four main directions of the development of the sonochemically driven fabrication of new composite/hybride materials:

- *SCS of nanocomposite structures based on simple and mixed metal oxides;*
- *Carbon-based nanocomposites (e.g. graphene, graphene oxide and carbon nanotubes);*
- *polymer-based hybrid nanostructures formation; and*
- *development of complex structures based on metal-organic frameworks.*

The examples cited in Table 4.2 of new composite/hybride materials prove that sonochemically driven technology has been successfully coping with the fabrication of efficient, eco-friendly adsorbents for the removal of organic dye and heavy metals from the water, bio- and electrochemical sensors, pseudocapacitive material for energy storage, polymer-based coating reinforced with nanofillers for the automotive or aircraft industry, etc. Further discussion will focus on US-assisted creation of the bio-functional and bioactive materials.

4.4 ULTRASOUND-ASSISTED CREATION OF THE BIO-FUNCTIONAL AND BIOACTIVE COMPOSITE STRUCTURES

4.4.1 A Few Words About Bioelectronics Devices

The movement of electronics technologies to the atomic scale and rapid advances in system, cell and molecular biology have recently resulted in significantly increased synergy between electronics and biology. Since we are talking about electronics, it is clear that we will talk about semiconductor structures and, first of all, about heterointerfaces between organic and inorganic matter. The remarkable features of semiconductor nanostructured materials, such as their unique quantum-mechanical properties, exceptionally high surface area and ease of chemical modification with biological ligands, have already been exploited for biomedical applications including imaging, biosensing and DNA analysis, cancer treatment and drug delivery (Tiwari et al. 2015). Further development of this area is directed towards the molecular functionalisation of semiconductor surfaces that is necessary to achieve progress in the chemical and biological detection by *biosensors* as well as to developing functional bioelectronics instruments, such as *implantable probes* and *neuroprosthetic devices.*

Among various biosensors, such as CMOS-transistor-based devices that have been used to monitor action potentials of excited neurons (Offenhäusser et al. 1997), or Si-nanowire

(SiNW)-based transistor read-outs of protein and DNA binding (Ingebrandt 2015), Si-based field-effect transistors (FETs) have attracted considerable attention because of their potential for miniaturisation, fast response time and easy integration with electronic manufacturing processes (Kaisti 2017). As shown schematically in Figure 4.8, an *FET biosensor* binds a charged analyte to receptor molecules (linker) on the gate surface to produce a change in conductance, resulting in a measurable electrical signal corresponding to the presence of said analyte. Surface functionalisation of silicon has initially focused on the functionalisation of OH-terminated silicon oxide surfaces, using silanisation, which strongly depends on the density of initial surface OH groups, the number of water molecules present and the temperature (Peng et al. 2015).

The use of the novel carbon materials, such as CNT or graphene, opens a very promising range of options for biosensors based on electronic devices made from these highly interesting materials (Kuila et al. 2011). These materials are used in FET devices as channel materials instead of bulk silicon in the traditional MOSFET structure. And one of the alternative methods of their surface functionalisation is a sonochemical approach.

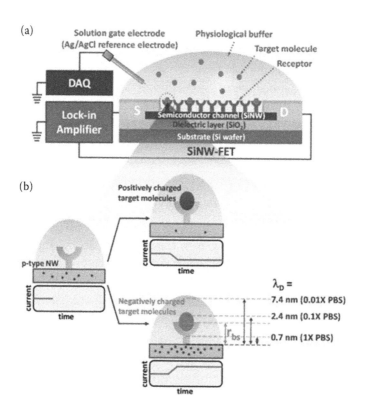

FIGURE 4.8   (a) Schematic illustrations of an experimental setup and the working principle of a SiNW-FET biosensor. A SiNW-FET contains a semiconductive channel, composed of a single SiNW or a bunch of SiNWs, which is electrically connected between the source and drain electrodes located on a Si wafer. (b) Receptor molecules are immobilised on the SiNW(s) to recognise specific targets with a SiNW-FET biosensor. (Adapted with permission from Li et al., 2014. Copyright 2014, The Royal Society of Chemistry.)

The examples of their sonochemical functionalisation can be found in Table 4.2. A non-hazardous sonochemical approach for *in situ* reduction and direct functionalisation of graphene oxide has been developed for nontoxic biomedical applications (Maktedar et al. 2017). The high cytocompatibility of functionalised with tryptamine GO confirms the low toxicity and an excellent biocompatibility of this composite. Conjugating CNTs with DNA, proteins, or carbohydrates is a general bio-functionalisation method which makes them able to create a new class of bioactive CNTs-based devices. However, MWCNTs are difficult to use directly for biological systems because of their hydrophobicity and insolubility. Here, sonochemistry comes to the aid – ultrasonication is used for uniform dispersion of CNTs in polymer nanocomposites by covalent functionalisation.

Not only functionalisation but biocompatibility plays a huge role for implantable probes and neuroprosthetic devices. Semiconductor nanomaterials (including nanocomposites), thanks to their unique optical, electrical and structural properties, may allow the creation of interfaces between neurons and the environment to restore or supplement the function of the nervous system lost during injury or disease (Fattahi et al. 2014). This provides new opportunities for implantable probes and neuroprosthetic devices that can be potentially included into brain-computer interface technologies. One of the first examples of hybrid bio-nanodevices, where absorption of light by thin films of quantum confined semiconductor nanoparticles of HgTe produced by the layer-by-layer assembly stimulating adherent neural cells via a sequence of photochemical and charge-transfer reactions, was reported in Pappas et al. (2007).

The field of the design of neural electrodes and their modification with electroactive materials, such as conducting polymers, CNTs, graphene or silicon nanowires, is also topical and promising (Agarwal et al. 2010; Parker et al. 2012). The need for the creation of composite/hybrid materials on their basis is conditional for the purpose of reducing the mechanical mismatch at the neural tissue interface as well as increasing the stability and biocompatibility of the electrode.

4.4.2 Biocompatible Composites

Creation of the biocompatible composite coatings is also important for bone repair. Novel composite hydrogels based on the combination of natural polymers, namely alginate and soy protein isolate, and bioactive glass particles, were developed by a sonochemical approach application (Silva et al. 2014). The bio-mineralisation process in simulated body fluid (SBF) was followed over time, and the results demonstrated that the composite materials have the ability to form a surface apatite layer after 7 days in SBF, and the design of novel composite hydrogels can be a suitable for bone regeneration. Another example of US-assisted manufacturing of polymer-based composite is reported in Mallakpour et al. (2018b). $CaCO_3$ NPs have been embedded into the tragacanth gum with the aim of biosorbent creation.

In the last few years, such composites as ceramics fillers in polymer matrices represent a new class of materials of high interest. The advantages of obtaining this kind of material by the method of sonochemical synthesis are shown in Ref. (Parra et al. 2009). Composites based on HDPE and PMMA with nanometric hydroxyapatite (HA), popular bone repairing materials for bone fixation joints, were obtained and studied. Nanometric HA particles are encapsulated into the polymeric PMMA matrix, but for the composites of HDPE, these interactions were not observed.

It should be noted that calcium hydroxyapatite ($Ca_{10}(PO_4)_3(OH)_2$) remains the most popular bioactive material widely used for hard tissue regeneration over the past 20 years due to its remarkable biocompatibility, high osteo-conductivity and close chemical similarity to biological apatite present in human hard tissues. A rapid, environmentally friendly and low-cost method to prepare HA NPs is proposed in Rouhani et al. (2010). In this method, hydroxyapatite is produced in a sonicated pseudo-body solution. All the time, researchers continue to work for improvement of these materials, since they have some disadvantages, including being fragile and antibacterial. The formation of HA-based nanocomposites with Ag antibacterial component incorporated within a bioactive stabilising component has the capacity to provide safe and effective antibacterial biomaterial. This approach has been applied to the development of antibacterial hydroxyapatite/silver (HA/Ag) materials (Vukomanovi et al. 2015). In addition, the incorporation of calcium silicate $CaSiO_3$ (CS) into HA was a useful approach to obtain composites with improved mechanical properties (Lin et al. 2011).

4.4.3 Silicon Surface Functionalisation based on Cavitation Processing

This section will present the method of the US-assisted formation of the nanostructured $Si/SiO_2/(CaSiO_3)$ composite demonstrating strong optical emission in the visible spectral range and biocompatibility confirmed by HA formation after storage in SBF solution (Savkina et al. 2016, 2017).

Among all semiconductors, Si remains as a key element in the area of the high-technology electronic devices fabrication. An effective way to expand its practice is developing new composite and hybrid Si-based systems joining diverse materials, which allows the integration of several key functions in a single structure and more complex properties for the new devices. For example, a device that combines a conventional silicon cell with a perovskite (such as $CH_3NH_3PbI_3$ or $HC(NH_2)_2]_{0.83}Cs_{0.17}Pb(I_{0.6}Br_{0.4})_3$) to increase the efficiency by converting more of the sun's energy into electricity has already been created. The efficiency of such devices can be increased up to 25% or even to 30%. Silicon-based hybrid organic–inorganic structures have also received most of the attention in the field of bio-interfacing.

Composite structures, which integrate the nanostructured silicon with bioactive silicates, are fabricated using the method of MHz sonication in the cryogenic environment (Savkina and Smirnov 2016). A MHz frequency Cryo Reactor with focused energy resonator was described in (Savkina and Smirnov 2010). US power was focused onto the sample surface in order to achieve high operation stability suitable for destructing sample surfaces. A sketch of the focusing setup is shown in Figure 4.9.

The cavitation processing of the silicon samples, providing extremely high impact pressures (about 8 bars in the focus), has resulted in the essential change of the surface morphology as well as optical and structural properties of silicon surfaces. After $15 \div 30$ min of sonication inside the structured region, nano- and subnanoscale objects (see Figure 4.10a–d),

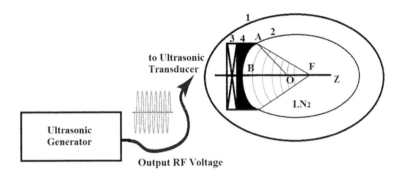

FIGURE 4.9   A schematic image of the Cryo Reactor: a stainless steel tank (1) with an internal reactor cavity (2) equipped with an acoustic system – a piezoelectric transducer (3) and a cylindrical copper lens (4) was used for the acoustic power enhancement; $AO = R_{cu}$ is a curvature radius, $AF = F$ is a focal distance, $\angle AFB = \gamma$ is the opening angle of the wave front. An oscillating voltage V_{US} applied to a piezoelectric ceramic (PZT-19) causes it to vibrate, thus delivering acoustic power into the liquid nitrogen- (LN2-) filled reactor cavity at US frequency of 3 MHz and 6 MHz (for two different resonators).

as well as dendrite-like objects (see Figure 4.10e,f), were revealed. The investigation of the chemical composition and optical properties of US-structured Si surfaces has indicated an essential oxidation of the samples after cavitation exposure. In particular, the weight per cent of the oxygen in the surface of none of the annealing samples has amounted to 12% (Savkina et al. 2015).

It was found that ultrasonically nanostructured regions demonstrate strong optical emission in the visible spectral range of 500–900 nm (see Figure 4.11).

Before annealing, a weak photoluminescence (PL) band in the energy region of 1.7 eV was obtained. The sonicated samples after annealing at 980°C (Figure 4.11, curve 2) in the atmospheric ambient have demonstrated PL bands with an intense peak around 565 nm (2.2 eV) and 750 nm (1.65 eV) (Savkina et al. 2017). Silicon samples, sonicated and annealing at 1100°C in an inert atmosphere, have demonstrated PL bands around 2.2 eV only. The described features could be related to the radiative processes in silicon oxide. PL bands in the energy range of 2.1–2.3 eV and around 1.7 eV are generally observed in Si-rich SiO_x ($x < 2$) matrices prepared by various techniques, such as plasma-enhanced chemical vapour deposition (PECVD) (Kenyon et al. 1996), implantation of silicon into silica (Shimizu-Iwayama et al. 1994) or thermal evaporation of silicon monoxide in a vacuum (Nesheva et al. 2002). At that, the properties of the PL bands appear to be very complex and strongly depend on the oxygen content in the SiO_x. Most researchers agree that PL bands at the higher energy (2.1–2.3 eV) are the result of defect luminescence, possibly from non-bridging oxygen centres or related oxygen vacancies. The lower energy PL band (around 1.7 eV) is associated with the radiative recombination of confined excitons.

In order to functionalise the silicon surface and to obtain a composite structure $Si/CaSiO_3$, the powder of gluconic acid calcium salt $C_{12}H_{22}CaO_{14}$ was added to the reactor vessel. Its heat decomposition during the cavitation processing leads to the formation of calcium oxide (CaO), carbon (C), carbon dioxide (CO_2) and water (H_2O) after sonication-obtained structures were annealed. The formation of a new phase on the silicon surface

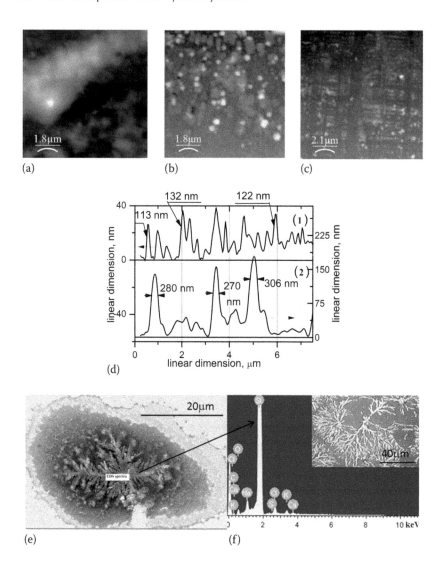

FIGURE 4.10 Typical AFM images of ultrasonically structured regions of Si samples: (a) – 3 MHz, (b) – 6 MHz, (c) – 6 MHz and at overlapping of the metal grid on the surface, (d) – a cross-sectional view of the structured surface after sonication at 6 MHz and at an overlapping of the metal grid on the surface (1), and after sonication at 6 MHz without a metal grid on the Si surface (2). (Adapted from Chapter 24 in *Springer Proceedings in Physics*, Structured silicon surface via cavitation processing for the photovoltaic and biomedical application, 183, 2016, 291–303, Savkina, R. K. and Smirnov, A. B., with permission. Copyright 2016, Springer Nature.) SEM micrograph (e) and the atomic composition (f) of the structured silicon surface exposed to the acoustic cavitation in liquid nitrogen at 6 MHz during 15 min. Inset: An optical image of the silicon surface exposed to the acoustic cavitation during 1 hour. (Reprinted from Kryshtab et al., *Mater Res Soc Symp Proc*, 1534, A87–A92. Copyright 2013, Cambridge University Press.)

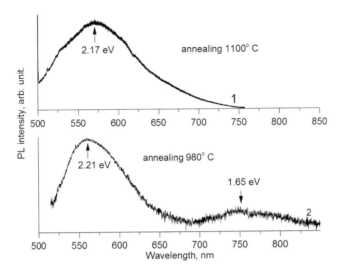

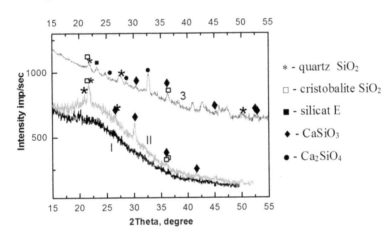

FIGURE 4.11 Photoluminescence spectra of Si surface sonicated (488 nm of Ar-Kr laser) with post-sonication annealing.

FIGURE 4.12 X-ray diffraction patterns of the typical Si samples at initial state and after MHz sonication (15 W/cm², 35 min) and annealing: I (black solid line) – the untreated silicon sample; II (red solid line) – silicon sample after sonication and annealing at 1100°C; III (blue solid line) – silicon sample after sonication and annealing at 980°C. (Adapted with permission from *Surf Coat Technol*, 343, Savkina et al., Silicon surface functionalization based on cavitation processing. Copyright 2017, Elsevier.)

was confirmed by the results of X-ray diffraction (Figure 4.12) and μ-Raman (Figure 4.13) spectroscopy investigations.

XRD patterns reveal a series of diffraction peaks which correspond to reflections from SiO_2, $CaSiO_3$ and Ca_2SiO_4 compounds. Polymorphs of SiO_2, such as quartz, cristobalite and stishovite, were found after sonication and annealing (see Figure 4.12). Calcium silicate ($CaSiO_3$) was formed with both orthorhombic and monoclinic crystal structures as well as with stable at ambient pressure triclinic crystal structures. Calcium orthosilicate

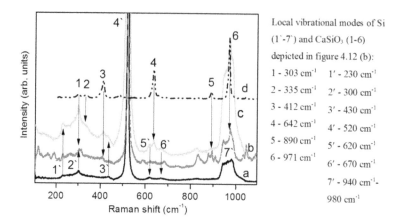

FIGURE 4.13   m-Raman spectra of the typical Si samples at initial state and after MHz sonication (15 W/cm^2, 35 min) and annealing: (a) – the untreated silicon sample; (b) – silicon sample after sonication and annealing at 1100°C; (c) – silicon sample after sonication and annealing at 980°C; (d) – Raman spectrum of wollastonite obtained from the RRUFF database (Lafuente et al. 2015). (Adapted with permission from *Surf Coat Technol*, 343, Savkina et al., Silicon surface functionalization based on cavitation processing. Copyright 2017, Elsevier.)

(Ca$_2$SiO$_4$) was formed with orthorhombic crystal structure after cavitation processing and subsequent annealing in the atmospheric ambient at 980°C for 1 hour only.

Raman spectra investigation has detected an appearance of the local vibrational-modes-characterised wollastonite form of CaSiO$_3$. The spectrum of wollastonite obtained from the RRUFF database (Lafuente et al. 2015) is presented in Figure 4.13 as reference. Ca-O local vibrational mode at about 336 cm^{-1} and 412 cm^{-1}, and Raman features at 642 cm^{-1} and 971 cm^{-1} corresponding respectively to the stretching vibration of the monomer SiO$_4$ and the stretching vibration of the chain SiO$_4$ tetrahedron are observed (Serghiou and Hammack 1993).

Calcium inosilicate (wollastonite, CaSiO$_3$) is a biomaterial with excellent bioactivity and biocompatibility. Currently, much research has been carried out to produce CaSiO$_3$ via the chemical precipitation method. The mechanochemical process is an alternative process route to prepare CaSiO$_3$ powder. This process exhibits several advantages, such as lower sintering temperature, homogeneous particle size with narrow particle size distribution and the formation of CaSiO$_3$ at ambient temperatures by using oxide materials, which are inert. The US-assisted method of the CS synthesis is not only much cheaper but also more environmentally friendly compared to the chemical precipitation or mechanochemical process. Using ultrasonic cavitation to manipulate semiconductor surfaces on a small scale allowed us to obtain silicon with a unique combination of light emission and biocompatibility.

The biocompatibility of the obtained composite structure was evaluated by examining bone-like apatite formation ability via SBF soaking method. Fourier transform infrared spectroscopy was performed on the silicon substrates before and after sonication and after storage in SBF (Figure 4.14). Initial transmission IR spectra exhibits strong IR absorption bands at 611 cm^{-1} assigned to a multiphonon absorption in the Si host as well as absorption associated with stretching (1107 cm^{-1}) and bending (817 cm^{-1}) vibration modes of the Si-O-Si bonds (Figure 4.14a). The position of the stretching vibration mode changes with thermal annealing from 1107 cm^{-1} to 1080 cm^{-1}. The vibration mode at 885 cm^{-1} can be associated with sub-oxidised silicon species (oxygen interstitials) and is due to the structural combination of Si_2O_3. This peak position is almost constant with different treatments.

Figure 4.14b shows the transmission IR spectra of processed samples for the spectral region from 2000 to 4500 cm^{-1}. The observed band around 2840 to 2960 cm^{-1} (see Figure 4.14b) is attributed to the C-H stretching modes (2846 cm^{-1} $v_{as}CH_2$, 2912 cm^{-1} v_sCH_2, 2955 cm^{-1} vCH_3) (Ahire et al. 2012). Thermal annealing of the sonicated Si substrates results in the appearance of the weak absorption band between 790 to 830 cm^{-1} (Figure 4.14a) and between 3250 to 3500 cm^{-1} (Figure 4.14b), which could be assigned to the N-H stretching modes (Shiohara et al. 2010). A strong absorption band around 1080 cm^{-1}, with a shoulder at 1200 cm^{-1}, could be connected with TO and LO SiO_2 optical phonon modes (Weldon et al. 1997) and points out the crystallisation of the oxidised silicon surface as a consequence of thermal annealing. It is necessary to note that the obtained functionalised layer exhibits antireflection properties in the MWIR spectral range.

The transmission IR spectra of processed samples normalised on initial spectral distribution of transmission for spectral regions from 400 to 1400 cm^{-1} are shown in Figure 4.15. Such a suitable approach has permitted the elimination of the contribution from the substrate in the initial state. The IR spectra of silicon samples stored monthly in SBF solutions demonstrate an appearance of the two optical bands associated with HA formation on the ultrasonically processed Si surface (depicted by Δ in Figure 14.5). The band around 550 cm^{-1} is assigned to HPO_4^{2-} and PO_4^{3-} groups, and characteristic peaks at 960 cm^{-1} and 1030 cm^{-1} correspond to the band of PO_4^{3-} asymmetric stretching modes in HA (Chakraborty et al.

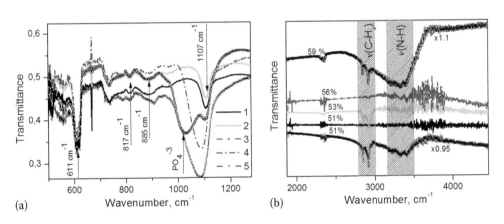

FIGURE 4.14 Transmission IR spectra of initial and ultrasonically processed Si samples: curve 1 – initial state; curve 2 – Si1 after cavitation processing; curve 3 – previous sample monthly stored in SBF solution; curve 4 – Si2 after cavitation processing with the following thermal annealing (at 1100°C for 2 hours); curve 5 – previous sample monthly stored in SBF solution. (Adapted with permission from *Surf Coat Technol*, 343, Savkina et al., Silicon surface functionalization based on cavitation processing. Copyright 2017, Elsevier.)

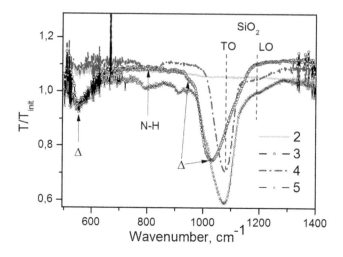

FIGURE 4.15 Transmission IR spectra of initial and ultrasonically processed Si samples normalised on initial spectral distribution: 1 – Tinit; curve 2 – Si1 after cavitation processing; curve 3 – previous sample monthly stored in SBF solution; curve 4 – Si2 after cavitation processing with the following thermal annealing (at 1100°C for 2 hours); curve 5 – previous sample monthly stored in SBF solution. Optical bands associated with HA formation depicted as Δ. (Adapted with permission from *Surf Coat Technol*, 343, Savkina et al., Silicon surface functionalization based on cavitation processing. Copyright 2017, Elsevier.)

2016). In a control sample of silicon that was not subjected to cavitation processing, there was no formation of hydroxyapatite.

Thus, the cavitation processing of the Si samples, besides a structuring of semiconductor surface and possible phase transformation of bulk material, results in the formation of functional oxide layers consisting of such materials as SiO_2, $CaSiO_3$ and Ca_2SiO_4. The obtained composite structure has demonstrated strong visible emissions in the spectral range of 500–800 nm as well as biocompatibility confirmed by the hydroxyapatite formation on the Si surface after storage in simulated body fluid solution. We believe that such composite structures can found applications in the spectral region of minimal tissue absorbance between 650 and 1350 nm, commonly known as the biological window, and can be used as biosensors or as components of bioelectronics implantable devices.

REFERENCES

Abbas, M., Torati, S. R. and Kim, C. G. 2015. A novel approach for the synthesis of ultrathin silica-coated iron oxide nanocubes decorated with silver nanodots ($Fe_3O_4/SiO_2/Ag$) and their superior catalytic reduction of 4-nitroaniline. *Nanoscale* 7:12192–204.

Abbas, M., Zhang, J., Lin, K. and Chen, J. 2018. Fe_3O_4 nanocubes assembled on RGO nanosheets: Ultrasound induced in-situ and eco-friendly synthesis, characterization and their excellent catalytic performance for the production of liquid fuel in Fischer-tropsch synthesis. *Ultrason Sonochem* 42:281–92.

Agarwal, S., Zhou, X., Ye, F., He, Q., Chen, G. C. K., Soo, J., Boey, F., Zhang, H. and Chen P. 2010. Interfacing live cells with nanocarbon substrates. *Langmuir* 26:2244–47.

Aghabeygi, S. and Khademi-Shamami, M. 2018. ZnO/ZrO_2 nanocomposite: Sonosynthesis, characterization and its application for wastewater treatment. *Ultrason Sonochem* 41:458–65.

Ahammad, A., Saleh, J., Lee, J.-J. and Rahman, M. A. 2009. Electrochemical sensors based on carbon nanotubes. *Sensors* 9:2289–2319.

Ahire, J. H., Wang, Q., Coxon, P. R., Malhotra, G., Brydson, R., Chen, R. and Chao, Y. 2012. Highly luminescent and nontoxic amine-capped nanoparticles from porous silicon: Synthesis and their use in biomedical imaging. *ACS Appl Mater Interfaces* 4:3285–92.

Bagheri, S., Aghaei, H., Ghaedi, M., Asfaram, A., Monajemi, M. and Bazrafshan, A. A. 2018. Synthesis of nanocomposites of iron oxide/gold (Fe_3O_4/Au) loaded on activated carbon and their application in water treatment by using sonochemistry: Optimization study. *Ultrason Sonochem* 41:279–87.

Bang, J. H. and Suslick, K. 2010. Applications of ultrasound to the synthesis of nanostructured materials. *Adv Mater* 22:1039–59.

Chakraborty, R., Sengupta, S., Saha, P., Das, K. and Das, S. 2016. Synthesis of calcium hydrogen-phosphate and hydroxyapatite coating on SS316 substrate through pulsed electrodeposition. *Mat Sci and Eng C* 69:875–83.

Chatel, G. 2017. *Sonochemistry: New Opportunities for Green Chemistry*. London, World Scientific Publishing.

Dezfuli, A. S., Ganjali, M. R. and Naderi, H. R. 2017. Anchoring samarium oxide nanoparticles on reduced graphene oxide for high-performance supercapacitor. *Appl Surf Sci* 402:245–53.

Dong, B., Huang, X., Yang, X., Li, G., Xia, L. and Chen, G. 2017. Rapid preparation of high electro-chemical performance $LiFePO_4$/C composite cathode material with an ultrasonic-intensified micro-impinging jetting reactor. *Ultrason Sonochem* 39:816–26.

Dubinsky, T. J., Cuevas, C., Dighe, M. K., Kolokythas, O. and Hwang, J. H. 2008. High-intensity focused ultrasound: Current potential and oncologic applications. *Am J Roentgenol* 190:191–99.

Eghbali-Arani, M., Sobhani-Nasab, A., Rahimi-Nasrabadi, M., Ahmadi, F. and Pourmasoud, S. 2018. Ultrasound-assisted synthesis of $YbVO_4$ nanostructure and $YbVO_4$/$CuWO_4$ nanocomposites for enhanced photocatalytic degradation of organic dyes under visible light. *Ultrason Sonochem* 43:120–35.

Fang, F. F. and Choi, H. J. 2008. Non-covalent self-assembly of carbon nanotube wrapped carbonyl iron particles and their magnetorheology. *J Appl Phys* 103:07A301-1–07A301-3.

Farghali, A. A., Khedr, M. H., El-Dek, S. I. and Megahed, A. E. 2018. Synthesis and multifunc-tionality of (CeO_2-NiO) nanocomposites via sonochemical technique. *Ultrason Sonochem* 42:556–66.

Fattahi, P., Yang, G., Kim, G. and Abidian, M. R. 2014. A review of organic and inorganic biomate-rials for neural interfaces. *Adv Mater* 26(12):1846–85.

Francony, A. and Petrier, C. 1996. Sonochemistry degradation of CCl_4 in aqueous solutions at two frequencies, 20 and 500 kHz. *Ultrason Sonochem* 3: S77–S82.

Ganguly, S., Das, P., Bose, M., Das, T. K., Mondal, S., Das, A. K. and Das, N. C. 2017. Sonochemical green reduction to prepare Ag nanoparticles decorated graphene sheets for catalytic perfor-mance and antibacterial application. *Ultrason Sonochem* 39:577–588.

Gedanken, A. 2004. Using sonochemistry for the fabrication of nanomaterials. *Ultrason Sonochem* 11:47–55.

Gharib, M., Safarifard, V. and Morsali, A. 2018. Ultrasound assisted synthesis of amide function-alized metal-organic framework for nitroaromatic sensing. *Ultrason Sonochem* 42:112–118.

Ghiyasiyan-Arani, M., Salavati-Niasari, M. and Naseh, S. 2017. Enhanced photodegradation of dye in waste water using iron vanadate nanocomposite; ultrasound-assisted preparation and characterization. *Ultrason Sonochem* 39:494–503.

Goyat, M. S., Jaglan, V., Tomar, V., Louchaert, G., Kumar, A., Kumar, K., Singla, A., Gupta, R., Bhan, U., Rai, S. K. and Sharma, S. 2017. Superior thermomechanical and wetting properties of ultrasonic dual mode mixing assisted epoxy-CNT nanocomposites. *High Perform Polym*. https://doi. org/10.1177/0954008317749021

Guiterez, M. and Henglein, A. 1990. Chemical action of ultrasound: Observation of an unprecedented intensity. *J Phys Chem* 94:3625–36.

Guo, Y., Yu, W., Chen, J., Wang, X., Gao, B. and Wang, G. 2017. Ag_3PO_4/rectorite nanocomposites: Ultrasound-assisted preparation, characterization and enhancement of stability and visible-light photocatalytic activity. *Ultrason Sonochem* 34:831–38.

Heidari, S., Haghighi, M. and Shabani, M. 2018. Ultrasound assisted dispersion of $Bi_2Sn_2O_7$-C_3N_4 nanophotocatalyst over various amount of zeolite Y for enhanced solar-light photocatalytic degradation of tetracycline in aqueous solution. *Ultrason Sonochem* 43:61–72.

Hinman, J. J. and Suslick, K. S. 2017. Nanostructured materials synthesis using ultrasound. *Top Curr Chem (Cham)* 375:12.

Hua, I., Hoechemer, R. H. and Hoffmann, M. R. 1995. Sonolytic hydrolysis of p-Nitrophenyl acetate: The role of supercritical water. *J Phys Chem* 99(8):2335–42.

Ingebrandt, S. 2015. Bioelectronics: Sensing beyond the limit. *Nat Nanotechnol* 10:734–735.

Jeon, W.-Y., Choi, Y.-B. and Kim, H.-H. 2017. Ultrasonic synthesis and characterization of poly(acrylamide)- co -poly(vinylimidazole)@MWCNTs composite for use as an electrochemical material. *Ultrason Sonochem* 43:73–79.

Jo, W.-K. and Yoo, H.-J. 2018. Combination of ultrasound-treated 2D g-C_3N_4 with Ag/black TiO_2 nanostructure for improved photocatalysis. *Ultrason Sonochem* 42:517–25.

Kaisti, M. 2017. Detection principles of biological and chemical FET sensors. *Biosensors and Bioelectronics* 98:437–448.

Kanthale, P., Ashokkumar, M. and Grieser, F. 2008. Sonoluminescence, sonochemistry (H_2O_2 yield) and bubble dynamics: Frequency and power effects. *Ultrason Sonochem* 15:143–50.

Kenyon, A. J., Trwoga, P. F., Pitt, C. W., and Rehm, G. 1996. The origin of photoluminescence from thin films of silicon-rich silica. *J Appl Phys* 79:9291–9300.

Karthik, N., Jebakumar, T. N., Edison, I., Sethuraman, M. G. and Lee, Y. R. 2017. Sonochemical fabrication of petal array-like copper/nickel oxide composite foam as a pseudocapacitive material for energy storage. *Appl Surf Sci* 396:1245–50.

Keramat, A. and Zare-Dorabei, R. 2017. Ultrasound-assisted dispersive magnetic solid phase extraction for preconcentration and determination of trace amount of Hg (II) ions from food samples and aqueous solution by magnetic graphene oxide (Fe_3O_4@GO/2-PTSC): Central composite design optimization. *Ultrason Sonochem* 38:421–29.

Khan, M., Tahir, M. N., Adil, S. F., Khan, H. U., Siddiqui, M. R. H., Al-warthan, A. A. and Tremel, W. 2015. Graphene based metal and metal oxide nanocomposites: synthesis, properties and their applications. *J Mater Chem A* 3:18753–18808.

Kryshtab, T. G., Savkina, R. K. and Smirnov, A. B. 2013. Nanoscale structuration of semiconductor surface induced by cavitation impact. *Mater Res Soc Symp Proc* 1534: A87–A92.

Kuila T., Bose, S., Khanra, P., Mishra, A. K., Kim, N. H. and Lee, J. H. 2011. Recent advances in graphene-based biosensors. *Biosens Bioelectron* 26:4637–48.

Kumar, P., Vellingiri, K., Kim, K.-H., Brown, R. J. C. and Manos, M. J. 2017. Modern progress in metal-organic frameworks and their composites for diverse applications. *Microporous and Mesoporous Materials* 253:251–65.

Kumar, A., Kumar, K., Ghosh, P. K. and Yadav, K. L. 2018. MWCNT/TiO_2 hybrid nano filler toward high-performance epoxy composite. *Ultrason Sonochem* 41:37–46.

Lafuente, B., Downs, R. T., Yang, H. and Stone, N. 2015. The power of databases: The RRUFF project. In *Highlights in Mineralogical Crystallography*, T. Armbruster and R. M. Danisi, eds. Berlin, Germany, W. De Gruyter, pp. 1–30.

Lee, J.-H., Rhee, K. Y. and Park, S. J. 2011. Silane modification of carbon nanotubes and its effects on the material properties of carbon/CNT/epoxy three-phase composites. *Compos A* 42:478–83.

Li, B.-R., Chen, C.-C., Kumar, U. R. and Chen, Y.-T. 2014. Advances in nanowire transistors for biological analysis and cellular investigation. *Analyst* 139:1589–1608.

Lin, K., Zhang, M., Zhai, W., Qu, H. and Chang, J. 2011. Fabrication and characterization of hydroxyapatite/ wollastonite composite bioceramics with controllable properties for hard tissue repair. *J Am Ceram Soc* 94:99–105.

Magdziarz, A. and Colmenares, J. C. 2017. In situ coupling of ultrasound to electro-and photo-deposition methods for materials synthesis. *Molecules* 22:216.

Mahboob, S., Haghighi, M. and Rahmani, F. 2017. Sonochemically preparation and characterization of bimetallic Ni-Co/Al_2O_3-ZrO_2 nanocatalyst: Effects of ultrasound irradiation time and power on catalytic properties and activity in dry reforming of CH_4. *Ultrason Sonochem* 38:38–49.

Maktedar, S. S., Mehetre, S. S., Avashthi, G. and Singh, M. 2017. *In situ* sonochemical reduction and direct functionalization of graphene oxide: A robust approach with thermal and biomedical applications. *Ultrason Sonochem* 34:67–77.

Mallakpour, S. and Behranvand, V. 2017. Sono-assisted preparation of bio-nanocomposite for removal of Pb^{2+} ions: Study of morphology, thermal and wettability properties. *Ultrason Sonochem* 39:872–82.

Mallakpour, S., Abdolmaleki, A. and Tabebordbar, H. 2018a. Employment of ultrasonic irradiation for production of poly(vinyl pyrrolidone)/modified alpha manganese dioxide nanocomposites: Morphology, thermal and optical characterization. *Ultrason Sonochem* 41:163–71.

Mallakpour, S., Abdolmaleki, A. and Tabesh, F. 2018b. Ultrasonic-assisted manufacturing of new hydrogel nanocomposite biosorbent containing calcium carbonate nanoparticles and tragacanth gum for removal of heavy metal. *Ultrason Sonochem* 41:572–81.

Mallakpour, S. and Mansourzadeh, S. 2018. Sonochemical synthesis of PVA/PVP blend nanocomposite containing modified CuO nanoparticles with vitamin B 1 and their antibacterial activity against Staphylococcus aureus and Escherichia coli. *Ultrason Sonochem* 43:91–100.

Mallakpour, S. and Javadpour, M. 2018. Sonochemical assisted synthesis and characterization of magnetic PET/Fe_3O_4, CA, AS nanocomposites: Morphology and physiochemical properties. *Ultrason Sonochem* 40:611–18.

Mallakpour, S., Abdolmaleki, A. and Azimi, F. 2017a. Ultrasonic-assisted biosurface modification of multi-walled carbon nanotubes with Thiamine and its influence on the properties of PVC/Tm-MWCNTs nanocomposite films. *Ultrason Sonochem* 39:589–96.

Mallakpour, S., Abdolmaleki, A. and Rostami, M. 2017b. Green synthesis of amino acid functionalized multiwalled carbon nanotubes/poly(amide-imide) based on NTrimellitylimido-S-valine nanocomposites by sonochemical technique. *J Polym Environ* 25:1–7.

Mallakpour, S. and Motirasoul, F. 2018. Ultrasonication synthesis of PVA/PVP/α-MnO_2-stearic acid blend nanocomposites for adsorbing Cd^{II} ion. *Ultrason Sonochem* 40:410–18.

Mallakpour, S. and Rashidimoghadam, S. 2018. Application of ultrasonic irradiation as a benign method for production of glycerol plasticized-starch/ascorbic acid functionalized MWCNTs nanocomposites: Investigation of methylene blue adsorption and electrical properties. *Ultrason Sonochem* 40:419–32.

Manickam, S. and Ashokkumar, M. 2014. *Cavitation. A Novel Energy-Efficient Technique for the Generation of Nanomaterials.* Singapore, Pan Stanford Publishing Pte. Ltd.

Margulis, M. A. 1992. Fundamental aspects of sonochemistry. *Ultrasonics* 30(3):152–55.

Martina, K., Tagliapietra, S., Barge, A. and Cravotto, G. 2016. Combined microwaves/ultrasound, a hybrid technology. *Top Curr Chem* (Z) 374:79.

Masjedi-Arani, M. and Salavati-Niasari, M. 2018. Cd_2SiO_4/graphene nanocomposite: Ultrasonic assisted synthesis, characterization and electrochemical hydrogen storage application. *Ultrason Sonochem* 43:136–45.

Mason, T. J. 1999. Sonochemistry: Current uses and future prospects in the chemical and processing industries. *Phil Trans R Soc Lond* A357:355–69.

Mason, T. J. and Lorimer, J. P. 2002. Introduction to applied ultrasonics. In *Applied Sonochemistry: Uses of Power Ultrasound in Chemistry and Processing*, Wiley-VCH Verlag GmbH & Co. KGaA, Weinheim, FRG.

Moghaddari, M., Yousefi, F., Ghaedi, M. and Dashtian, K. 2018. A simple approach for the sono-chemical loading of Au, Ag and Pd nanoparticle on functionalized MWCNT and subsequent dispersion studies for removal of organic dyes: Artificial neural network and response surface methodology studies. *Ultrason Sonochem* 42:422–33.

Mohamed, H. H. 2017. Sonochemical synthesis of ZnO hollow microstructure/reduced graphene oxide for enhanced sunlight photocatalytic degradation of organic pollutants. *J Photochem Photobiol A Chem* 353:401–8.

Mohammadinezhad, A., Marandi, G. B., Farsadrooh, M. and Javadian, H. 2017. Synthesis of poly(acrylamide-co-itaconic acid)/MWCNTs superabsorbent hydrogel nanocomposite by ultrasound-assisted technique: Swelling behavior and Pb (II) adsorption capacity. *Ultrason Sonochem*. https://doi.org/10.1016/j.ultsonch.2017.12.028

Mosleh, S., Rahimi, M. R., Ghaedi, M., Dashtian, K. and Hajati, S. 2018. Sonochemical-assisted synthesis of CuO/Cu$_2$O/Cu nanoparticles as efficient photocatalyst for simultaneous degradation of pollutant dyes in rotating packed bed reactor: LED illumination and central composite design optimization. *Ultrason Sonochem* 40(Pt A):601–10.

Naderi, H. R., Ganjali, M. R., Dezfuli, A. S. and Norouzi, P. 2016. Sonochemical preparation of a ytterbium oxide reduced graphene oxide nanocomposite for supercapacitors with enhanced capacitive performance. *RSC Adv* 6:51211–220.

Nesheva, D., Raptis, C. and Perakis, A. 2002. Raman scattering and photoluminescence from Si nanoparticles in annealed SiO$_x$ thin films. *J Appl Phys* 92:4678–83.

Nikitenko, S. I. and Pflieger, R. 2017. Toward a new paradigm for sonochemistry: Short review on nonequilibrium plasma observations by means of MBSL spectroscopy in aqueous solutions. *Ultrason Sonochem* 35B:623–30.

Offenhäusser, A., Maelicke, C., Matsuzawa, M. and Knoll, W. 1997. Field-effect transistor array for monitoring electrical activity from mammalian neurons in culture. *Biosens Bioelectron* 12:819–26.

Pang, W., Rupich, S. M., Shafiq, N., Gartstein, Y. N., Malko, A. V. and Chabal, Y. J. 2015. Silicon surface modification and characterization for emergent photovoltaic applications based on energy transfer. *Chem Rev* 115:12764–96.

Pappas, T. C., Wickramanyake, W. M. S., Jan, E., Motamedi, M., Brodwick, M. and Kotov, N. A. 2007. Nanoscale engineering of a cellular interface with semiconductor nanoparticle films for photoelectric stimulation of neurons. *Nano Letters* 7:513–19.

Parker, C. B., Raut, A. S., Brown, B., Stoner, B. R. and Glass, J. T. 2012. Three-dimensional arrays of graphenated carbon nanotubes. *J Mat Res* 27:1046–53

Parra, C., González, G. and Albano, C. 2009. Synthesis and characterization of composite materials HDPE/HA and PMMA/HA prepared by sonochemistry. *Macromol Symp* 286:60–9.

Poddar, M. K., Arjmand, M., Sundararaj, U. and Moholkar, V. S. 2018. Ultrasound-assisted synthesis and characterization of magnetite nanoparticles and poly(methyl methacrylate)/magnetite nanocomposites. *Ultrason Sonochem* 43:38–51.

Poinot, T., Benyahia, K., Govin, A., Jeanmaire, T. and Grosseau, P. 2013. Use of ultrasonic degradation to study the molecular weight influence of polymeric admixtures for mortars. *Construction and Building Materials* 47:1046–52.

Ramandi, S., Entezari, M. and Ghows, N. 2017. Sono-synthesis of novel magnetic nanocomposite (Ba-α-Bi$_2$O$_3$-γ-Fe$_2$O$_3$) for the solar mineralization of amoxicillin in an aqueous solution. *Phys Chem Res* 5(2):253–68.

Rivas, D. F. and Kuhn, S. 2016. Synergy of microfluidics and ultrasound process intensification challenges and opportunities. *Top Curr Chem* (Z) 374:70.

Rouhani, P., Taghavinia, N. and Rouhani, S. 2010. Rapid growth of hydroxyapatite nanoparticles using ultrasonic irradiation. *Ultrason Sonochem* 17:853–56.

Sander, J. R. G., Zeiger, B. W. and Suslick, K. S. 2014. Sonocrystallization and sonofragmentation. *Ultrason Sonochem* 21(6):1908–15.

Savkina, R. K. and Smirnov, A. B. 2010. Nitrogen incorporation into GaAs lattice as a result of the surface cavitation effect. *J Phys D: Appl Phys* 43(42):425301–7.

Savkina, R. K., Smirnov, A. B., Kryshtab, T. and Kryvko, A. 2015. Sonosynthesis of microstructures array for semiconductor photovoltaics. *Mater Sci Semicond Process* 37:179–84.

Savkina, R. K., Gudymenko, A. I., Kladko, V. P., Korchovyi, A. A., Nikolenko, A. S., Smirnov, A. B., Stara, T. R. and Strelchuk, V. V. 2016. Silicon substrate strained and structured via cavitation effect for photovoltaic and biomedical application. *Nanoscale Res Lett* 11:183.

Savkina, R. K. and Smirnov, A. B. Structured silicon surface via cavitation processing for the photovoltaic and biomedical application, Chapter 24 in *Springer Proceedings in Physics*, 2016, Vol. 183, pp. 291–303. ISBN:978-3-319-30736-7.

Savkina, R. K., Smirnov, A. B., Gudymenko, A. I., Morozhenko, V. A., Nikolenko, A. S., Smoliy, M. I. and Kryshtab, T. G. 2017. Silicon surface functionalization based on cavitation processing. *Surf Coat Technol* 343:17–23.

Serghiou, G. C. and Hammack, W. S. 1993. Pressure induced amorphization of wollastonite (CaSiO₃) at room temperature. *J Chem Phys* 98:9830–34.

Shamskar R. F., Meshkani, F. and Rezaei, M. 2017. Ultrasound assisted co-precipitation synthesis and catalytic performance of mesoporous nanocrystalline NiO-Al₂O₃ powders. *Ultrason Sonochem* 34:436–47.

Shimizu-Iwayama, T. and Fujita, K. 1994. Visible photoluminescence in Si⁺-implanted silica glass. *J Appl Phys* 75:7779–83.

Silva, R., Bulut, B., Roether, J., Kaschta, J., Schubert, D. W. and Boccaccini, A. R. 2014. Sonochemical processing and characterization of composite materials based on soy protein and alginate containing micron-sized bioactive glass particles. *J Mol Struct* 1073:87–96.

Shah, N., Bhangaonkar, K., Pinjari, D. V. and Mhaske, S. T. 2017. Ultrasound and conventional synthesis of CeO₂/ZnO nanocomposites and their application in the photocatalytic degradation of rhodamine B Dye. *J Adv Nanomaterials* 2:133–145.

Shchukin D. G., Radziuk, D. and Möhwald, H. 2010. Ultrasonic fabrication of metallic nanomaterials and nanoalloys. *Annu Rev Mater Res* 40(1):345–62.

Shende, T. P., Bhanvase, B. A., Rathod, A. P., Pinjari, D. V. and Sonawane, S. H. 2018. Sonochemical synthesis of Graphene-Ce-TiO₂ and Graphene-Fe-TiO₂ ternary hybrid photocatalyst nanocomposite and its application in degradation of crystal violet dye. *Ultrason Sonochem* 41:582–9.

Shiohara, A., Hanada, S., Prabakar, S., Fujioka, K., Lim, T. H., Yamamoto, K., Northcote, P. T. and Tilley, R. D. 2010. Chemical reactions on surface molecules attached to silicon quantum dots. *J Am Chem Soc* 132:248–253.

Skorb, E. V., Baidukova, O., Andreeva, O. A., Cherepanov, P. V. and Andreeva, D. V. 2013. Formation of polypyrrole/metal hybrid interfacial layer with self-regulation functions via ultrasonication. *Bioinspired, Biomimetic and Nanobiomater* 2:123–9.

Skorb, E. V. and Möhwald, H. 2016. Ultrasonic approach for surface nanostructuring. *Ultrason Sonochem* 29:589–603.

Skrabalak, S. E. 2009. Ultrasound-assisted synthesis of carbon materials. *Phys Chem Chem Phys* 11(25):4930–42.

Sonawane, S. H., Bhanvase, B. A., Kulkarni, R. D. and Khanna, P. K. 2014. Ultrasonic processing for synthesis of nanocomposite via in situ emulsion polymerization and their applications. In *Cavitation. A Novel Energy-Efficient Technique for the Generation of Nanomaterials*, ed. S. Manickam and M. Ashokkumar, 301–341. Boca Raton, FL, CRC Press/Taylor & Francis Group.

Song, Y., Li, Y., Li, J., Li, Y., Niu, S. and Li, N. 2018. Ultrasonic-microwave assisted synthesis of three-dimensional polyvinyl alcohol carbonate/graphene oxide sponge and studies of surface resistivity and thermal stability. *Ultrason Sonochem* 42:665–671.

Sponer, J. 1990. Dependence of the cavitation threshold on the ultrasonic frequency. *Czech J Phys* 40(10):1123–32.

Suh, W. H., Jang, A. R., Suh, Y.-H. and Suslick, K. S. 2006. Porous, hollow, and ball-in-ball metal oxide microspheres: preparation, endocytosis, and cytotoxicity. *Adv Mate* 18:1832–37.

Suslick, K. S., McNamara, W. B. III and Didenko, Y. 1999. Hot spot conditions during multi-bubble cavitation. In *Sonochemistry and Sonoluminescence*, ed. L. A. Crum, T. J. Mason, J. Reisse, K. S. Suslick, 191–204. Dordrecht, The Netherlands, Kluwer Publishers.

Tanhaei, M., Mahjoub, A. R. and Safarifard, V. 2018. Sonochemical synthesis of amide-functionalized metal-organic framework/graphene oxide nanocomposite for the adsorption of methylene blue from aqueous solution. *Ultrason Sonochem* 41:189–195.

Tiwari, A., Patra, H. K. and Turner A. P. F. 2015. *Advanced Bioelectronic Materials*. Hoboken, New Jersey, John Wiley & Sons and Salem, Massachusetts, Scrivener Publishing LLC.

Valange, S., Chatel, G., Amaniampong, P. N., Behling, R. and Jérôme, F. 2018. Ultrasound-assisted synthesis of nanostructured oxide materials: Basic concepts and applications to energy. In *Advanced Solid Catalysts for Renewable Energy Production*, eds. S. González-Cortés & F. E. Imbert, 177–215. Hershey, PA, IGI Global Book Series: Advances in Chemical and Materials Engineering (ACME).

Vichare, N. P., Senthilkumar, P., Moholkar, V. S., Gogate, P. R. and Pandit, A. B. 2000. Energy analysis in acoustic cavitation. *Ind Eng Chem Res* 39:1480–6.

Vinoth, R., Karthik, P., Devan, K., Neppolian, B. and Ashokkumar, M. 2017. TiO_2-NiO p-n nano-composite with enhanced sonophotocatalytic activity under diffused sunlight. *Ultrason Sonochem* 35(Pt B):655–63.

Vukomanovi, M., Repnik, U., Zavašnik-Bergant, T., Kostanjšek, R., Skapin, S. D. and Suvorov, D. 2015. Is nano-silver safe within bioactive hydroxyapatite composites? *ACS Biomater Sci Eng.* https://doi.org/10.1021/acsbiomaterials.5b00170

Weldon M. K., Stefanov, B. B., Raghavachari, K. and Cha-bal, Y. J. 1997. Initial H_2O-induced oxidation of Si (100). *Phys Rev Lett* 79:2851–4.

Wood, R. J., Lee, J. and Bussemaker, M. J. 2017. A parametric review of sonochemistry: Control and augmentation of sonochemical activity in aqueous solutions. *Ultrason Sonochem* 38:351–370.

Xu, H., Zeiger, B. W. and Suslick, K. S. 2013. Sonochemical synthesis of nanomaterials. *Chem Soc Rev* 42:2555–2567.

Zak, K., Majid, W. H. A., Wang, H. Z., Yousefi, R., Golsheikh, A. M. and Ren, Z. F. 2013. Sonochemical synthesis of hierarchical ZnO nanostructures. *Ultrason Sonochem* 20:395–400.

Zhang, K., Park, B.-J., Fang, F.-F. and Choi, H. J. 2009. Sonochemical preparation of polymer nanocomposites. *Molecules* 14:2095–2110.

Zhao, S., Chen, D., Wei, F., Chen, N., Liang, Z. and Luo, Y. 2017. Removal of Congo red dye from aqueous solution with nickel-based metal-organic framework/graphene oxide composites prepared by ultrasonic wave-assisted ball milling. *Ultrason Sonochem* 39:845–52.

Zhu, Z., Garcia-Gancedo, L., Flewitt, A. J., Xie, H., Moussy, F. and Milne, W. I. 2012. A critical review of glucose biosensors based on carbon nanomaterials: Carbon nanotubes and graphene. *Sensors* 12:5996–6022.

Zinatloo-Ajabshir, S., Mortazavi-Derazkola, S. and Salavati-Niasari, M. 2017. Simple sonochemical synthesis of Ho_2O_3-SiO_2 nanocomposites as an effective photocatalyst for degradation and removal of organic contaminant. *Ultrason Sonochem* 39:452–60.

Zinatloo-Ajabshir, S., Mortazavi-Derazkola, S. and Salavati-Niasari, M. 2018. Nd_2O_3-SiO_2 nanocomposites: A simple sonochemical preparation, characterization and photocatalytic activity *Ultrason Sonochem* 42:171–82.

CHAPTER 5

Solid State Composites Produced by Ion Beam Processing

Aleksej Smirnov

CONTENTS

5.1 INTRODUCTION: LITHOGRAPHY-BASED METHODS VERSUS ION BEAM PROCESSING FOR NANOELECTRONICS FABRICATION

Since 2000, industrial microelectronics have overcome the boundary design standards of 100 nm and thus were transformed into nanoelectronics. Until now, the mass production of ultra large scale integration (ULSI) (Simon et al. 1986) with sub-100 nm topological standards was provided by short-wave UV lithography (UV – ultraviolet, wavelength of 193 nm). To achieve a resolution of 45 nm and less, it is necessary to significantly complicate the lithographic process using phase correcting templates, immersion mode exposure, double patterning and the like (Flagello 2007). Currently, the world's leading manufacturers

(Intel, United States) have achieved spatial resolution in critical elements (gate structures of integrated transistors) of 32 nm at design norm 10 nm given by ITRS (international technology roadmap for semiconductors) (Kahng et al. 2015). Figure 5.1 shows technologies which are available for nanoelectronics manufacturing as well as development and research fields (Technology and manufacturing day of INTEL 2017). Progress in the field of nanotechnology can be represented by transistor density (MTr/mm^2) improvements; see Figure 5.2 (Technology and manufacturing day of INTEL 2017).

As a result, modern industrial lithographic equipment is the most expensive part of a complex of technological equipment for the production of ULSI equipment providing minimum topological dimensions. Scaling of integrated devices in the region of less than 32 nm has shown that existing optical lithographs are unable to cope with this task without reducing the wavelength in the EUV (extreme ultraviolet) region to 13.5 nm.

The physical problems encountered in the development and creation of such equipment affect all aspects of the creation of a lithographic machine: from the development of powerful reliable sources of radiation to an optical system lithographer, and up to the need to create new templates and resistors. These problems led to the fact that the EUV-lithograph was manufactured only in the form of two experimental machines from ASML Inc. (Holland) EUV Alpha Demo Tool,* which were installed in the research centres of CNSE (Albany, USA) and IMEC (Leuven, Belgium), and which were constantly perfected and finalised. Several more systems (NXE-3100 and NXE-3300B) were manufactured. The final, industrialised version was completed in late 2017 thanks to a new design of the high-power CO_2 laser system that generates EUV radiation in the lithography tool's source chamber when it blasts a series of tin droplets.

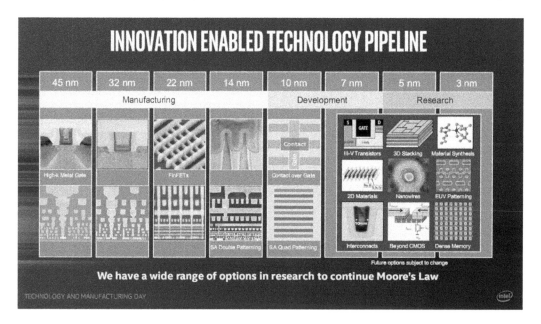

FIGURE 5.1   Technology and manufacturing day of INTEL

---

* https://www.asml.com/euv-is-at-the-cusp-of-being-introduced-in-volume-chip-production-the-industrialization-metrics-of-euv-most-importantly-productivity-and-availability-will-drive-the-decision-/en/s41905?rid=41906

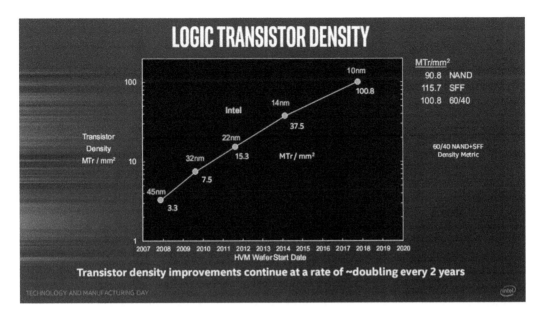

FIGURE 5.2   Transistor density improvements continue at a rate of doubling every two years.

Considering alternative technologies of sub-100 nm structures formation, it is neces-sary to pay attention to the possibilities of electron-beam lithography (EBL) (Smith and Suzuki 2007) and nanoimprint technology (Viheriälä et al. 2007). Physical principles of electron-beam lithography consist in the modification of the properties of electron-sensitive resist film by focused electron beams and the subsequent chemical removing of the exposed resist. Recently, research in the field of electron-lithographic methods has intensified and, under laboratory conditions, the formation of individual nano-objects up to 5 nm was achieved (Bernstein et al. 2004). At the same time, the potential of molecular nanoelectron-ics (Vieu et al. 2000) and single-electronics (Henini et al. 1999) methods of the multi-beam EL lithography were demonstrated. In parallel with the studies mentioned above, the capa-bilities of EBL for use in small-scale production of nanoelectronics products are enhanced.

However, it is impossible to completely solve the problem of EBL for the formation of sub-100 nm elements. There are a number of factors that limit the resolution of the method. In particular, this is a known effect associated with scattering and reflection of the elec-tron beams which expose the resist. Another problem, limiting the use of EL lithography for the formation of nanostructures by direct exposure of the resist, is the low stability electron-sensitive resistances to chemically active plasma used to transfer the mask pat-tern to the functional underlying layer. This often leads to the necessity of using electronic lithography two- and three-layer masks (Tseng et al. 2003), which significantly compli-cate the lithographic process. Another drawback of EBL is the low productivity and, as a result, low efficiency for large-scale industrial production. An alternative to UV- and EUV-lithography is *technology self-organisation of nanostructures during ion beam processing*, which allows the formation of chaotic and ordered nanostructures directly on the surface of a semiconductor wafer or creates nanomasks for subsequent alloying, as, for example, in Smirnov et al. (2003).

Ion implantation demonstrates one of the best examples of the technique, producing a precise dose of impurity as well as a uniform and shallow junction for semiconductor devices manufacturing (Rimini 1995). Besides, ion implantation is a very powerful tool for synthesis and modification of the solid state low-dimensional and nanoscale composite structures (Rizza and Ridgway 2016a). In contrast to EBL, the alternative technology of ion beam processing makes it possible to form arrays of nanostructures (e.g. Si nanowires) simultaneously on the entire surface of the semiconductor wafer. Application of the technology of nanostructures' self-organisation, in combination with EBL, can lead to an increase in the productivity of the lithographic process and does not require the use of expensive lithographic equipment.

5.2 ION IMPLANTATION APPROACH TO THE SOLID STATE STRUCTURES FORMATION BY THE SURFACE NANOSTRUCTURING AND NANOSTRUCTURES SELF-ORGANISATION

We do not present a complete literature review describing the ion beam processing approach to the new material fabrication since several excellent articles and book chapters already exist (e.g. Mazzoldi et al. 2005; Nastasi and Mayer 2006; Bernas 2010; Krasheninnikov and Nordlund 2010). Further discussion will be devoted to some obvious successes in the field of interaction between ion beams and solids – surface nanostructuring and self-organisation of nanostructures in the ion irradiated matrix.

5.2.1 Surface Nanostructuring

The ion implantation method allows the obtainment of new micro- and nanostructured materials with a wide range of topological features on the surface. Significant efforts of the scientific community were applied to the study and interpretation of the microscopic dynamics of the process of structuring the surface of metals, insulators and basic semiconductors (Ge, Si) under the action of ion bombardment. Such implantation parameters as ion energy, fluence and angle of the ion beam incidence are significant to optimise the evolution of nanostructures. In particular, the length scales of nanoripples are found to be mainly dependent on ion energy (Chini et al. 2002). Depending on the energy level, stationary waves with a certain orientation (ions of $10^2 \div 10^5$ eV), terraces, nanotubes and nanodots may form on the semiconductor target ("nanodots"/"nanoholes" at ions energy up to 10^2 eV). The results presented in Firestein et al. (2016) have demonstrated that by changing Ag ion energy in the range of 2–20 kV (acceleration voltage), it is possible to selectively fabricate an Ag/BN nanocomposite with crystalline or amorphous structures.

Typically, surface formations are characterised by a variation of the field of height in intervals $(0.1 \div 1)$ μm with characteristic dimensions from tens of angstroms to several hundreds of nanometres. The morphology and kinetics of structure formation are determined by the interaction of the ion flux angle with the target surface. It was shown that normal-incidence (NI) ion beams can result in the formation of nanoscale objects on the surface of elemental Si (Gago et al. 2001; Ziberi et al. 2009), Ge (Ziberi et al. 2009; Wei et al. 2009) and compound semiconductor GaSb (Facsko et al. 1999; Plantevin et al. 2007). At the same time, well-ordered hexagonal arrays of InP nanodots (Frost et al. 2000) and well-aligned ripple structures on the surface of a single crystal of 3C-SiC (Jiaming et al. 2008) were created by oblique-incidence (OI) ion bombardment. It should also be noted that ripples formed by OI processing of binary compounds can have a much higher degree of order than those formed by bombardment of elemental materials (Motta et al. 2012). Important factors in this matter are the substrate temperature and the fact of the target rotation. Figure 5.3 shows AFM and SEM images of nanometre scale relief after ion beam processing of Si (Figure 5.3a,b), GaSb (Figure 5.3c) and InSb (Figure 5.3d) substrates.

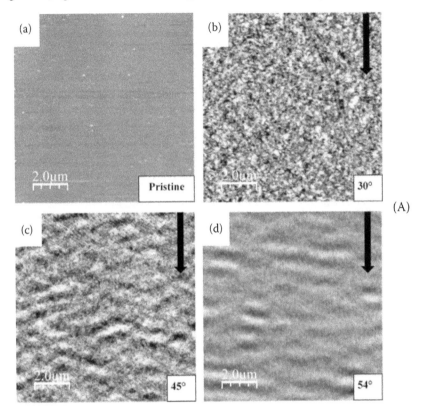

FIGURE 5.3 Pattern evolution on semiconductor surface as a function of angle of incidence, fluence, type and temperature of substrate: (A) AFM images (10μm×10μm) of silicon before (a) and after 60 keV Ar+-ion irradiation (fluence: 2×10^{18} ions cm^{-2}) (b–d) as a function of angle of incidence, θ. (Reprinted with permission from *Appl Surf Sci*, 310, Garg et al., 60 keV Ar+-ion induced pattern formation on Si surface: Roles of sputter erosion and atomic redistribution, 147–153, Copyright 2014, Elsevier); (B) AFM image of Si (1 0 0) surface irradiated at θ = 60° for fluences mentioned in the figures A–D. Black arrows on the images show the direction of incident ion beam. (Reprinted with permission from *Appl Surf Sci*, 310, Garg et al., 60 keV Ar+-ion induced pattern formation on Si surface: Roles of sputter erosion and atomic redistribution, 147–153, Copyright 2014, Elsevier); (C) SEM images of the tilted views of modified GaSb (001) surfaces bombarded with a 3 keV Ar$^+$ ion beam (J~2.98×10^{13} ion/s cm^2, ϕ~2.15×10^{17} ion/cm^2) at normal incidence, presented for increasing sample temperatures. (Reprinted with permission from *Vacuum*, 66, Thompson, M. W., Atomic collision cascades in solids, 99–114, Copyright 2018, Elsevier) (D) SEM images of the tilted views (52° off the normal) of modified InSb (001) surfaces bombarded with a 3 keV Ar$^+$ ion beam (J~6.3×10^{13} ion/s cm^2, ϕ~4.5×10^{17} ion/cm^2) at normal incidence, presented for increasing sample temperatures. (Reprinted with permission from *Vacuum*, 66, Thompson, M. W., Atomic collision cascades in solids, 99–114, Copyright 2018, Elsevier.)

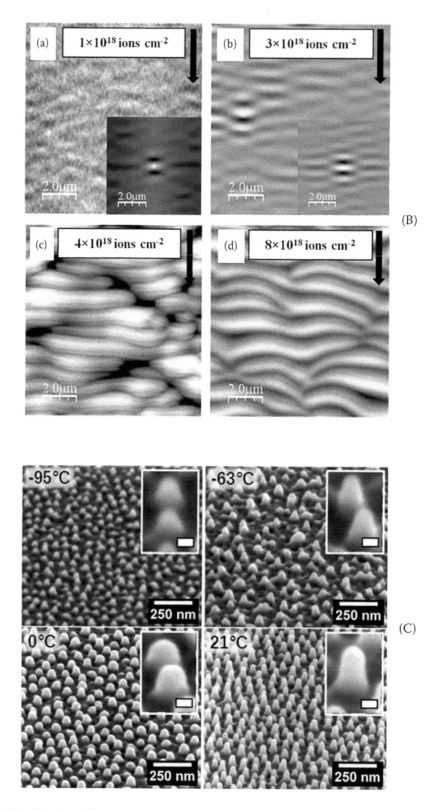

FIGURE 5.3   (Continued)

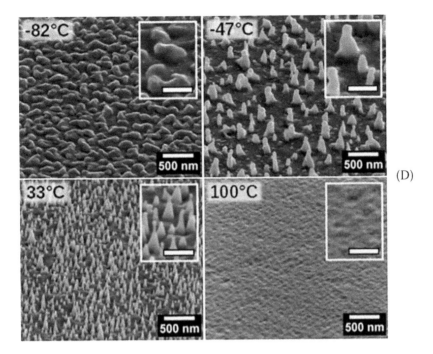

FIGURE 5.3 (Continued)

It is established that nanoscale ripples are formed in a certain range of angles of incidence ions. Their width is determined by the energy and type of the primary beam. It should also be noted that the values of these quantities are different for each set of experimental conditions (type of primary ions, their energy and angle of incidence). For example, it was experimentally established that the depth (or dose of irradiation) is the lowest when using nitrogen ions for Si irradiation. It is one order of magnitude smaller than when Si is irradiated by ions O^{2+} and is two orders of magnitude smaller when using ions Ar^+ (Smirnov et al. 1999).

The formation of relief begins only at a certain dose of primary ions (the degree of amorphisation of the modified layer), and the wavelength depends on the thickness of the modified layer. At a qualitative level, the modified layer can be considered frozen liquid, which flows only in the region of linear cascades under the action of the volume force created, like the cascades themselves, by primary ions flying at an angle to the surface (Norris 2012).

There are several mathematical models that adequately describe the development processes of mesoscopic topography and the formation of microscopic (Bradley-Harper model) and nanometre scale (nonlocal equation of erosion) relief during processing by ion beam, which allows solving this task without an expensive experiment. Most researchers believe that the formation of nanoscale ripples is due to only erosion-related phenomena, but the accumulated experimental facts show that the process of shaping the undulating terrain is probably two stages and consists of the stage of formation of the embryonic relief and the subsequent development of surface topography during ion sputtering surface. The mechanism of formation of the embryonic relief is obviously connected with the formation of the modified (amorphous) layer and development in its "hydrodynamic" instability.

Most commonly, surface nanostructuration driven by the ion beam is understood as the interplay between surface re-organisation processes such as roughening and smoothening due to erosion and diffusion. The most well-known deterministic model is the equation Bradley-Harper (Bradley and Harper 1988), within the framework of which the formation of nanoscale ripples is explained by the curvature dependence of the sputtering yield. This equation was intended, first of all, to explain the development of nanoscale ripples, which according to the existing erosion model is

$$\frac{\partial \theta}{\partial t} = -\frac{J}{\rho}\cos^2\theta \frac{\partial}{\partial x}\left(\frac{Y(\theta_0 - \theta)\cos(\theta_0 - \theta)}{\cos\theta}\right),$$

where:

θ = the slope angle between the local normal to the surface and the z axis
θ_0 = angle of bombardment
J = the ion flux density
ρ = density of target atoms
Y = the sputtering coefficient

As already noted, The Bradley-Harper model is local, i.e. the coordinates of the points of incidence of the primary ion and the yield of the secondary ion in it are considered to be coincident. In Bradley and Harper (1988), a three-dimensional erosion model was achieved, which allowed the explanation of the formation of surface topography induced by ion bombardment both for incidence angles close to the normal and when the ion beam is close to grazing incidence.

Figure 5.4 presents greyscale plots for the nontrivial stable steady-state solutions (Eq. 16 in Bradley and Shipman [2012]) of the equations showing the change of the surface relief amplitude (Eqs. 11 and 12 in Bradley and Shipman [2012]). It can be seen that the calculated surface relief is similar to that obtained experimentally (see Figure 5.3).

An alternative approach is based on the theory of stress relaxation developed by Cuerno and Moreno-Barrado et al. (2015) and Norris (2012). In particular, Cuerno, with co-authors,

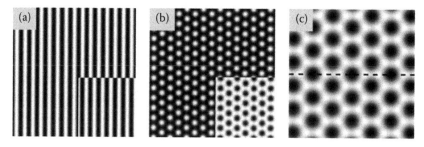

FIGURE 5.4   Greyscale plots for the nontrivial stable steady-state solutions of the amplitude equations showing the change of the surface relief amplitude (Eqs. 11 and 12 in Bradley and Shipman [2012]). Panel (a) shows the surface ripple. Panel (b) is a greyscale plot of the surface amplitude for a hexagonal array of nanodots. In Panel (c), a hexagonal array of nanoholes is shown. (Reprinted with permission from *Appl Surf Sci*, 258, Bradley, R. M. and Shipman, P. D., A surface layer of altered composition can play a key role in nanoscale pattern formation induced by ion bombardment, 4161–4170, Copyright 2012, Elsevier.)

shows that nonuniform generation of stress across the damaged amorphous layer induced by their irradiation is a key factor behind the range of experimental observations. For the convenience of practical use of the results of mathematical modelling, they can be presented in the form of software packages, allowing one to calculate the area of existence of the required type of relief depending on the ion energy and flux density, angle of bombardment, sputtering time, etc.

The first recorded observation of the sputtering process belongs to V. Grove (1852), who drew attention to the disintegration of cathodes in glow discharge tubes. It was established that this process is caused by the impact of high-energy ions on the cathode (Baragiola 2004). There are many theories and models explaining the main peculiarities of the sputtering process, but at present it is the most widespread theory, developed by Thompson (1981, 1987, 2002). If the energy of the ion colliding with the surface is sufficient to overcome the repulsive potential barrier created by surface atoms, it will penetrate into the solid state after its first encounter with the surface atom is rejected. It will continue to move in a solid state, losing energy in subsequent collisions and deviations, until it finally arrives in a state of rest. The energy transferred to atoms of a solid in the process inhibition may be large enough to knock them out of position equilibrium. At each collision, a vacancy and an atom can be formed upon impact. If the recoil atoms have a sufficient energy reserve, then they will cause further displacements of atoms, called the collision cascade. In this case, the number of recoil atoms will increase and their energy will gradually decrease. In the final analysis, the energy of recoil atoms is itself dissipated in the collision processes (see figure 1 in Ref. Castro et al. 2012).

In the linear cascade regime, the collision region is approximately ellipsoidal (Bradley and Harper 1988; Makeev et al. 2002). Surfaces of the same energy release are ellipsoids of revolution with a major axis, coinciding with the direction of the motion of ions, the centres of which are located below the surface at a depth that increases with increasing ion energy. As a result of this energy distribution, the atoms on the surface, the energy received and the momentum directed to the surface are sufficient for overcoming the binding energy and will be ejected or sputtered from the surface. They will be distributed over energy, emission direction and the number of atoms per incident ion because of the statistical nature of collisional processes (Makeev et al. 2002).

It is possible to determine the average sputtering coefficient Y, which will be linearly dependent on the rate of energy release in elastic collisions near the surface. It will be inversely proportional to the binding energy of the target atoms with the surface. In addition, since the axis of the average collisional cascade coincides with the direction of ion incidence, the energy released near the surface will depend on the angle of the bombardment $Y(\theta)$; hence, the sputtering coefficient.

At a normal incidence of ions, the region of the intersection of the ellipsoidal cascade volume with the surface increases so that the coefficient sputtering also increases. However, if the angle of incidence is further, then the recoil atoms formed near the surface will leave it without generation of higher generations of recoil atoms. In this way, there is a reduction in the totality of the recoil atoms and a decrease in sputtering ratio. Finally, as the angle of incidence of the ions approaches sliding angle, the number of reflected ions increases and the coefficient sputtering tends towards zero. This behaviour of the sputtering coefficient Y is borrowed from the experimental work on sputtering with low-energy (1.05 keV) ions (Ochesner 1975).

5.2.2 Low-Dimensional and Composite Structures Obtained by Ion Implantation: 'Inclusions in Dielectric Matrix'

Ion implantation of one or more low-solubility species into a suitable matrix, accompanied or followed by thermal annealing, plays a special role in the nanocomposite fabrication with novel optoelectronic and magnetic properties (Meldrum et al. 2001). These are metallic precipitates (e.g. Au or Ag) or semiconductor nanocrystals embedded in a ceramic matrix such as SiO_2, Al_2O_3 or another for nonlinear optics, luminescence, optical switching applications, etc. It is also about ferromagnetic composite structures obtained by ion implantation.

It was demonstrated that separating the nucleation and growth of nanoparticles by multiple implantation steps can substantially reduce the breadth of the size distribution of nanoparticles formed by ion implantation (using Au implanted SiO_2 as a model system Ramaswamy et al. 2005]). Direct synthesis of nanoparticles formed by dual implantation of large $((4-10)\times10^{16}\,cm^{-2})$ and equal doses of Cd+S, Zn+Te, Cd+Te or Pb+Te ions into SiO_2 substrate was studied by Desnica et al (2004). An isotropically distributed 3D ensemble of these nanoparticles in the amorphous SiO_2 matrix, as well as nanocrystals of compound semiconductors CdS, ZnTe or CdTe, appearing after high-temperature annealing, were detected (see Raman spectra in Figure 5.5). It was found that at high ion doses, a fraction of implanted atoms synthesised already during implantation into amorphous aggregates of compound semiconductor, which transform into crystalline nanoparticles after annealing.

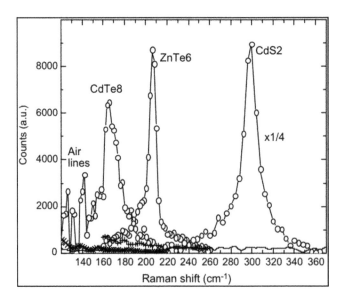

FIGURE 5.5 Raman spectra of CdS, ZnTe and CdTe semiconductor compound formed in SiO_2 matrix after implantation and high-temperature annealing. (Reprinted with permission from *Nucl Instrum Methods Phys Res A*, 216, Desnica et al., Direct ion beam synthesis of II-VI nanocrystals, 407–413, Copyright 2004, Elsevier.)

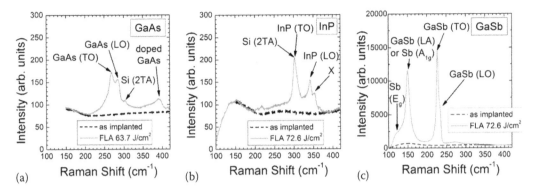

FIGURE 5.6 Raman spectra of GaAs (a), InP (b) and GaSb (c) semiconductor nanocrystals formed by ion implantation and sub-second FLA in a $SiO_2/Si/SiO_2$ layer stack on Si. The black and red curves depict the spectra of the as-implanted and annealed samples for each compound, respectively. (Reprinted with permission from Wutzler et al., Liquid-phase epitaxy of binary III-V nanocrystals in thin Si layers triggered by ion implantation and flash lamp annealing. *J Appl Phys* 117:175307. Copyright 2015, AIP Publishing LLC.)

A similar approach makes it possible to form III-V compound semiconductor nanocrystals, namely, GaAs, GaSb and InP, by ion implantation and sub-second flash lamp annealing (FLA) in a $SiO_2/Si/SiO_2$ layer stack on Si (Wutzler et al. 2015). The III-V nanocrystals formation takes place during the cooling down phase after an intense light pulse has melted the III-V implanted a-Si layer (see Raman spectra in Figure 5.6). During this cooling phase, small Si seeds recrystallise first, acting as templates for the epitaxial regrowth of the III-V nanocrystals. The higher segregation coefficients of V group elements in Si compared to that of III group elements lead to an excess of III group elements in the cooling melt, which in turn induces the formation of metallic group III precipitates.

The advantages and the perspectives of the ion implantation technique in the formation of metal or metal alloys nanoclusters in the dielectric matrix have been presented in detail by Mazzoldi and Mattei (2005). It should also be noted that the ion-shaping technique has emerged as a powerful tool to engineer real three-dimensional architectures in the form of embedded nanostructures with tuneable morphology and spatial orientation and has been proven to be particularly well-adapted for sculpting metal-glass nanocomposites (Rizza and Ridgway 2016b).

A strategy to design and fabricate composite metallic-dielectric substrates for optical spectroscopy and imaging has been proposed (Carles et al. 2011). Different architectures consisting of three-dimensional (3D) patterns of metallic nanoparticles embedded in dielectric layers were fabricated by low-energy ion beam synthesis (10 keV/1×10^{16} cm^{-2}–20 keV/1.5×10^{16} cm^{-2}) and consist of a delta-layer of Ag-nanoparticles with controlled size, density and location, embedded in the silica/silicon matrix (Benzo et al. 2013) and in the silicon nitride matrix (Bayle et al. 2015). It was shown that using the SiN_x matrix as a diffusion barrier for Ag allows the introduction of a higher amount of Ag in the matrix, compared to SiO_2. Hence, the formation of a high density of Ag-nanoparticles during the implantation process leads under particular implantation conditions to an in-plane self-organisation effect. The fabrication of such embedded 2D super-crystals should be of great interest for modulating acoustic or thermodynamic properties in the THz regime.

Examples of the successful application of the ion implantation for the formation of nanocomposites and thin films of metal oxide nanoheterostructures are presented by Romanyuk et al. (2007) and Singh et al. (2012). In Ref. (Romanyuk et al. 2007), results of the irradiation of SiO_2/Si structure with Cu^+ ions (energies $(40 \div 50)$ keV and doses $(10^{14} \div 10^{15})$ cm^{-2}) using the acoustic loading of the target are presented. The implantation of such a target in the acoustic wave field is accompanied by metal nanoclusters formation, a decrease in the deposition threshold and an increase of the copper diffusion. The irradiation of thin Ag films by silver ions of 100 MeV energy is proposed by authors Singh et al. (2012) to form Ag^+ nanoparticles at high concentrations $(10^{15} \div 10^{18}$ cm$^{-3})$ on the surface of a metal target. The high-quality nanocrystalline coatings for the flash memory devices based on C_{70}, C_{60}/Si (Tiwari et al. 1995), SiO_2 and $Si/CeO_2/SiO_2$ (Boer et al. 2001) were manufactured by the Si ion implantation with an energy of 35 keV and a dose of 4.3×10^{16} cm^{-2}. The nanocrystalline size was $(2 \div 6)$ nm. Belov and co-authors (2010) successfully localised the distribution of carbon ion runs within the thickness of SiO film with doses $(6.0 \div 10^{16})$ and $(1.2 \div 10^{17})$ cm^{-2} with an energy of 40 keV and produced light-emitting nanocomposites of "white" light on base $Si(C^+)$ (Belov et al. 2010).

Thus, surface nanostructuring and self-organisation of embedded nanocrystals elaborated by ion implantation has already been determined. However, the formulating of the general mechanisms of these phenomena faces certain difficulties and requires research.

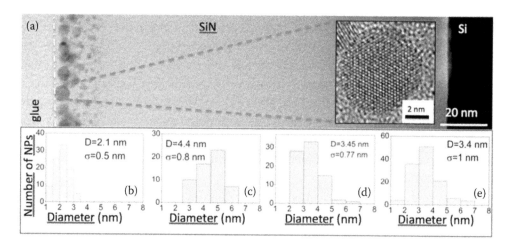

FIGURE 5.7 The Bright Field XS-TEM image of the layer implanted with the intermediate dose $(3 \times 10^{16}$ ions cm$^2)$. Size distributions of the samples implanted at 20keV for doses (b) $1.5 \times 1 \times 10^{16}$ cm^{-2}, (c) $3 \times 1 \times 10^{16}$ cm^{-2} and (d) $4.5 \times 1 \times 10^{16}$ cm^{-2}; the size histogram for the sample implanted at 8 keV for $9 \times 1 \times 10^{15}$ cm^{-2} is added in (e). In inset is shown an HREM image of an isolated Ag nanocrystal. (Reprinted with permission from Bayle et al., Ag doped silicon nitride composites for embedded plasmonics. *Appl Phys Lett* 107:101907. Copyright 2015, AIP.)

5.3 APPLICATIONS

Electronic and optical properties of low-dimensional and composite structures obtained during implantation are different from those of bulk materials and might find technological application for nanophotonics and nanoscale magnetism. It was shown that post-growth processing with cold, high fluence, Fe implantation was key in producing InGaAsP-based THz devices with good emitter characteristics (Fekecs et al. 2011). The implantation method for large-scale synthesis of high-quality grapheme films (Garaj et al. 2010; Zhang et al. 2013) and for InSbN layer formation of nitrogen incorporated into InSb wafer (Wang et al. 2012) was demonstrated. Ion implantation is used to locally modify the solid surface to create periodic plasmonic microstructures with metal nanoparticles (Bayle et al. 2015; Stepanov et al. 2013). Another example deals with a low-energy implantation (25 keV) of thin (Ga, Mn) As layers with a very low fluence of either O or Ne ions with completely suppressed ferromagnetism and which could be applied as a method for tailoring nanostructures in the layers (Yastrubchak et al. 2010).

Ion implantation of magnetic species into ferroelectric single crystal targets as a radically novel approach to prepare film nanoparticulate magnetic-metal ferroelectric oxide composites are presented in Alguero et al. (2014). It was shown that ion implantation of a high dose (10^{17} ions cm^{-2}) of magnetic Co^+ into ferroelectric $BaTiO_3$ crystals resulted in cobalt nanoclusters, directly observed by HR-TEM, embedded in an amorphised oxide matrix at the subsurface. Co^+ implantation of $Pb(Mg_{1/3}Nb_{2/3})O_3$-$PbTiO_3$ single crystals showed the formation of an ensemble of ferromagnetic nanoparticles embedded in an amorphised layer at the surface of the ferroelectric crystal (Torres et al. 2011). These materials are an alternative to multiferroic oxide epitaxial columnar nanostructures whose magnetoelectric response is far from expectations.

It should also be noted that the ion beam processing not only creates but also modifies the material properties. For example, the formation of nanowires on the surface of GaN is usually carried out with normal irradiation of the target by Ga^+ ions (Dhara et al. 2005), and on the surface of Si samples with B^+, P^+ and As^+ ions with energies 50 keV (Kanungo et al. 2009).

In the case of composited systems containing two or more different materials, the most obvious example is ion mixing and alloying during ion irradiation, which can occur at their interfaces. For example, the irradiation stability of amorphous SiOC and crystalline Fe interface was investigated by *in situ* 1MeV Kr ion mixing (Su et al. 2016). The microstructure evolutions of the Fe/SiOC interface after irradiation at 50, 298, 423 and 573 K with doses of 4.5×10^{15} ions/cm^2 are present as Figure 5.8b–e, respectively. The most important feature observed from these data is less intermixing and higher radiation stability at elevated temperatures. This is in contrast with metallic and metal-semiconductor systems with negative mixing heat where the amount of mixing increases with irradiation temperature in the temperature-dependent mixing regime. This result can be used to aid in the potential design of amorphous ceramic/metal composites for service in extreme irradiation environments.

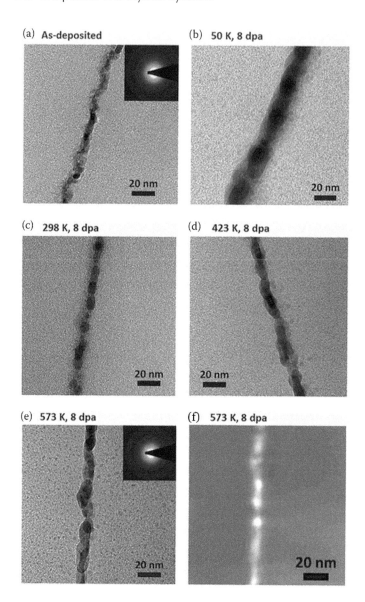

FIGURE 5.8 Typical cross-sectional bright field TEM micrographs of marker specimen (a) before and after (b) 50 K, (c) 298 K, (d) 423 K, (e) 573 K irradiation. (f) The scanning transmission electron microscopy (STEM) image of marker specimen after 573 K irradiation. (Reproduced from Su et al., *Mater Res Lett*, 4, 198–203 (2016) under the terms of the Creative Commons Attribution License.)

The effectiveness of ion implantation as a controlled means of achieving n- or p-doping in suspended single-layer graphene (Nitrogen and Boron doping) was demonstrated in Kepaptsoglou et al. (2015), thus paving the way to practical implementations of intrinsic graphene-based FETs.

In the next section, formation of nanoscale patterns on the surface of ternary compound HgCdTe (MCT), as well as fabrication of the composite structure integrating MCT with metal-oxide inclusions obtained by ion implantation for THz application, will be presented.

5.4 COMPOSITE STRUCTURE INTEGRATING TERNARY COMPOUND (HgCdTe) WITH METAL-OXIDE (Ag₂O) INCLUSIONS OBTAINED BY ION IMPLANTATION FOR THz APPLICATION

It is known that MCT is a unique semiconductor solid solution for photon detectors from near-wavelength to long-wavelength infrared regions (from NWIR to LWIR) (Rogalsky 2011). The ability to detect sub-terahertz radiation by MCT-based structures was also discussed (Dobrovolsky et al. 2008; Kryshtab et al. 2016). Ion implantation of MCT structures is a commonly used method for the fabrication of the IR devices. The advantages of this technique in producing a uniform and shallow junction are recognised. An implant, getting into the epitaxial layer, initiates an active restructuring of the defect structure of MCT, which changes the epilayer carrier type. As a result, n-on-p (boron-implanted) (Holander-Gleixner et al. 1997) and p-on-n (arsenic-implanted) (Mollard et al. 2009) photodiodes are fabricated. At the same time, it is well known that ion implantation induces mechanical stress in MCT layers, which has been exploited to improve their electrical and optical properties. It was shown that implantation-induced stress is an important factor influencing the depth of p-n junctions in MCT-based structures (Ebe et al. 1999).

Materials used in this study were p-$Hg_{1-x}Cd_xTe$ (x ~0.223) epilayers grown on [111]-oriented semi-insulating $Cd_{1-y}Zn_yTe$ (y = 0.04) substrates from a Te-rich solution at 450°C by liquid-phase epitaxy. Specimens 1×1 cm in size were cut for measurements from wafers. An optical microscope was used to determine the thickness of grown layers, which was about 17 μm. MCT layers were implanted with the boron B^+ and silver Ag^+ ions. The implantation has been carried out by means of a Vezuvii-5 implanter, allowing this to work in the energy range from 100 up to 140 keV with normal and oblique-incidence ion bombardment.

It should be noted that the p-n junction is produced in MCT by ion bombardment with beam fluence ranging from 10^{13} ions/cm² to 10^{15} ions/cm² and ion energy ranging from 100 keV to 300 keV. At that, the appearance of the extended defects in addition to the point defects occurs (Williams et al. 1997). But, a low dose (~10^{13} cm⁻²) was selected since the near-surface region of the MCT target becomes saturated with point defects in such conditions. In particular, there are vacancies and mercury interstitial sites.

The mathematical simulation of the process of ion implantation with the use of the software package TRIM_2008 allowed the parameters of the radiation-induced disordering region of MCT to be determined. The energy and dose of implanted ions as well as the parameters of the MCT region subjected to the ion-induced disordering are presented in Table 5.1.

5.4.1 Surface Topometry Investigation

It was found that the ion bombardment of the samples investigated has resulted in the essential change of the physical and structural properties of the MCT surface. In the range of nanoscale, arrays of holes and mounds are generated on a (111) MCT surface as a result of the normal incident ions' bombardment. An SEM image of periodic height modulations induced on a (111) MCT surface by 100 keV B^+ is shown in Figure 5.9a. Arrays of nanoscale

TABLE 5.1 Parameters of the MCT Region Subjected to the Ion-Induced Disordering

Type of Implanted Ions and Ion Beam Incidence Angle θ (°)	B⁺,0	Ag⁺,0	Ag⁺,0	Ag⁺,45
Energy of implanted ions (keV)	100	100	140	140
Dose of implanted ions (cm⁻²)	3×10^{13}	3×10^{13}	4.8×10^{13}	4.8×10^{13}
Projected range Rp (μm)	0.22	0.0365	0.045	0.04
Straggling ΔRp (μm)	0.17	0.024	0.028	0.0186
Maximal mechanical stresses σ(Pa)	1.4×10^{3}	2.2×10^{5}	2.9×10^{5}	1.28×10^{5}
Maximum doping Cmax (m⁻³)	9.5×10^{23}	5×10^{24}	5.25×10^{24}	3.38×10^{24}
Vacancy concentration Cv (N/Å ion)	20	4.5	4.67	4.68
The coefficient of crystal lattice contraction β(m⁻³)	3.51×10^{-31}	1.25×10^{-32}	1.25×10^{-32}	1.25×10^{-32}

Source: Reprinted with permission from Smirnov, A. B. and Savkina, R. K., *Nanophysics, Nanomaterials, Interface Studies, and Applications*, Springer, Cham. Copyright 2017, Springer Nature.

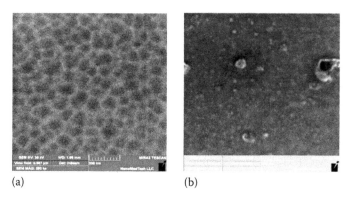

(a)   (b)

FIGURE 5.9 Typical SEM images of a (111) MCT surface generated as a result of the normal incident ions bombardment with boron (a) and silver (b) ions with 100 keV.(Reprinted with permission from *Nanophysics, Nanomaterials, Interface Studies, and Applications,* Nanostructuring surfaces of HgCdTe by ion bombardment, 195, 2017, Smirnov, A. B. and Savkina, R. K., Copyright 2017, Springer.)

mounds depicted in Figure 5.9b are generated on the (111) MCT surface as a result of 100 eV Ag+ ions implantation.

AFM results of the initial specimen topometry (Figure 5.10a) show that the surface plane is densely and regularly packed with round grains with preferred size of 25 nm in diameter. This means that the studied epitaxial film is characterised by a considerable nonequilibrium resource. As a rule, this state is concentrated in mechanical stresses of the local character (grains-pores), which is confirmed by the presence of a network of quasipores 3.5–10 nm in depth and 50–160 nm in diameter. The root-mean-square roughness (RMS) parameter for 1×1-μm² initial surface fragments was in the energy range of 2.45–3.34 nm.

After the double implantation with silver ions at NI regime (Figure 5.10f), the MCT surface undergoes an essential transformation. The RMS parameter is reduced to 2.17 nm from the initial value of 3.34 nm. The quasipores are almost unobservable, the grain boundaries are strongly smeared and some grains, keeping their sizes variable, form chains with channels between them.

Figure 5.10b shows the AFM reconstruction of periodic height modulations ("nanohole" pattern) induced on a MCT (111) surface with 100 keV NI B+ ions processing. After low-temperature annealing, the MCT surface became denser. The study of the microhardness with the use of a Shimadzu HMV-2000 pointed to an increase of its value to 12%. The ordered grid of quasipores is not observed (see Figure 5.10c). At the same time, some grains become consolidated.

NI Ag+ ion bombardment gives rise to the emergence of a uniform array of nano-islands 5–25 nm in height and with a base diameter of 13–35 nm (see Figure 5.10d). With increasing fluence from 3×10^{13} cm^{-2} to 4.8×10^{13} cm^{-2} and at OI regime, the diameter of nano-islands is found to increase, and the conjoined structure has been formed as shown in Figure 5.10e. The size distribution histogram for nano-objects on the specimen surface irradiated at the OI regime has a linear section (on the logarithmic scale). The presence of such a section means that the structures with fractal geometry are formed on the surface of a semiconductor compound epitaxial film. The corresponding 2D-fast Fourier transformation (FFT) has been depicted in insets on Figure 5.10. They reveal that there is no signature of the order of nanostructures over the surfaces for all regimes.

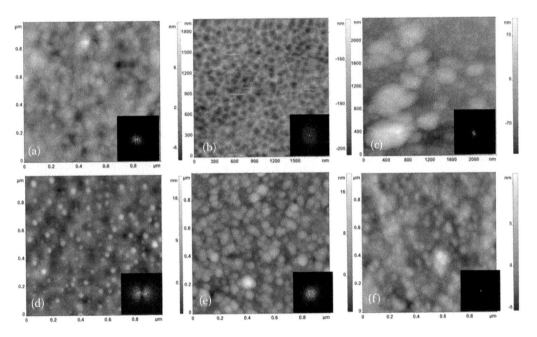

FIGURE 5.10   AFM images of a (111) MCT surface generated as a result of the ions bombardment with B+ and Ag+ ions: (a) – typical virgin surface, (b) – B+ ($\theta = 0°$, 100 keV, 3×10^{13} cm^{-2}), (c) – B+ ($\theta = 0°$ and post-implantation annealing, 100 keV, 3×10^{13} cm^{-2}), (d) – Ag+ ($\theta = 0°$, 100 keV, 3×10^{13} cm^{-2}), (e) – Ag+ ($\theta = 45°$, 140 keV, 4.8×10^{13} cm^{-2}), (f) – Ag+ ($\theta = 0°$, 140 keV, two doses of 4.8×10^{13} cm^{-2}). Inset: Fourier transforms of AFM images. (Reprinted with permission from *Nanophysics, Nanomaterials, Interface Studies, and Applications,* Nanostructuring surfaces of HgCdTe by ion bombardment, 195, 2017, Smirnov, A. B. and Savkina, R. K., Copyright 2017, Springer Nature.)

Thus, it was found that in the nanoscale range, arrays of holes and mounds on the (111) MCT surface have been fabricated using 100–140 keV B^+ and Ag^+ ions beam irradiation. Under certain irradiation conditions, in particular after NI irradiation with Ag^+ ions, a uniform array of nano-islands 5–25 nm in height was obtained. The topometry investigation after OI irradiation with Ag^+ ions points to the nanostructures with fractal geometry.

5.4.2 XRD Investigation

X-ray rocking curves (RC) obtained in the coherent-scattering region from the symmetrical $\omega/2\theta$-scanning for MCT-based structures before and after implantation point to the compression of the boron-implanted MCT and tension of the silver-implanted MCT layers (Smirnov et al. 2014). It was found that mechanical strain induced in MCT ternary compounds under ion beam influence is responsible for the evolution of surface morphology as well as stipulates peculiarity of the mass and charge transport in this material (Smirnov et al. 2013). X-ray diffraction measurements have confirmed the formation of a new phase in the subsurface region of MCT after silver implantation – cubic Ag_2O (Smirnov et al. 2017). Such composite structure integrating the nanostructured ternary compound (HgCdTe) with metal-oxide (Ag_2O) inclusions demonstrates an inductive-type impedance (or "negative capacitance") and is characterised by the extended region of the spectral sensitivity in comparison with the basic material (MCT).

Photoelectron spectra were performed to investigate the chemical state of MCT-based structures after ion bombardment. The survey scan of MCT layers implanted by Ag^+ was recorded, and the peaks of Hg, Cd, Te, Ag, along with O and C, were identified. The deconvolution of investigated XPS spectra shows clearly that the sample's surface is oxidised. The binding energies of Te photoelectrons were estimated to be 572.70 ± 0.15 eV. The second peak is observed at 576.80 ± 0.15 eV and can be connected with Te-based oxide. The cadmium $3d$ spectrum also indicates partial oxidation. The XPS signal clearly indicates the existence of silver in the surface region: Ag 3d5/2 centred at 367.70 ± 0.15 eV; a satellite at 369.20 ± 0.15 eV; and minor fraction of second satellite at 371.40 ± 0.15 eV, suggesting with high probability the existence of Ag_2O and AgO phases at the MCT surface.

The structural properties of the MCT epilayers were also investigated using X-ray high-resolution reciprocal space mapping (HR-RSM). The micro-defects system in the initial material is apparently compensated, which is confirmed by the symmetric form of the initial RC and RSM. The observed distribution of the intensity along axis q_z indicates the existence in the initial material of some structural heterogeneity caused by the existence of the vacancies ($q_z<0$) and interstitials ($q_z>0$) (Figure 5.11a). Implantation with boron results in insignificant changes in the structural properties of MCT. RSMs of the intensity distribution around (111) sites for the MCT layer implanted by silver ions have shown the following results, depicted in Figure 5.11b–d.

The formation of the significant halo in the central part of the map of the implanted sample (Figure 5.11b–d) indicates the increase in the MCT film imperfection after the implantation performed. After processing both with NI (Figure 5.11c) and with OI (Figure 5.11d) geometry, the significant increase in the half-width of the ω-scan (q_x direction) in the central part of the RSMs is observed. It can mean the formation of misorientation-type

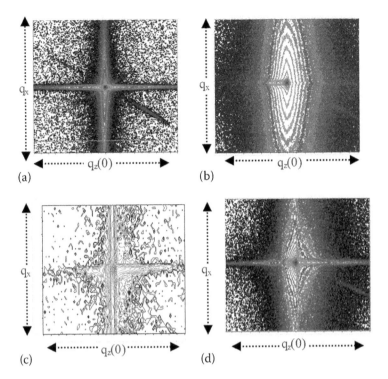

FIGURE 5.11 Reciprocal space maps of MCT-based heterostructure obtained from the combined ω and (ω–2θ)-scans using high-resolution modules: (a) – before implantation, (b) – Ag^+ (θ = 0°, 140 keV, two doses of 4.8×10^{13} cm^{-2}), (c) – Ag^+ (θ = 0°, 100 keV, 3×10^{13} cm^{-2}), (d) – Ag^+ (θ = 45°, 140 keV, 4.8×10^{13} cm^{-2}). (Reprinted with permission from *Nanophysics, Nanomaterials, Interface Studies, and Applications*, Nanostructuring surfaces of HgCdTe by ion bombardment, 195, 2017, Smirnov, A. B. and Savkina, R. K., Copyright 2017, Springer Nature.)

defects as a result of the ion bombardment. Moreover, the appearance of asymmetry in the q_z direction points to the initiation of the tensile deformation in the samples investigated after Ag^+ implantation.

It should also be noted that peculiarity at $q_z < 0$ on RSM of MCT-based heterostructure (Figure 5.11d) can mean the formation of the new phase. XRD investigation in the grazing incidence (GI) scheme has confirmed the formation of the new phase after NI and OI silver implantation (Smirnov et al. 2017). The results of X-ray diffraction analysis of MCT films with the composition 0.23 irradiated with 100 keV silver ions proved the formation of polycrystalline MCT layer with x=0.2 on the surface (ICDD PDF 00-051-1122) and (111) Ag_2O phase (ICDD PDF 00-041-1104) in the near-surface region (see Figure 5.12). The GI diffractograms have been collected by irradiating the samples at an incident angle (θ_{inc}) of 1°. The penetration depth at this incidence angle was ~400 nm.

5.4.3 A Small Discussion about Nanostructuring of MCT Surface and Composite Structure Ag_2O-MCT for THz Application

Thus, the experimental study demonstrates the nanostructurisation as well as change of the chemical composition on the MCT surface after NI and OI ion beam irradiation.

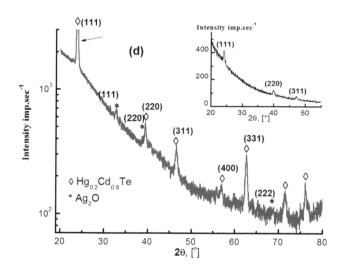

FIGURE 5.12   GI XRD spectrum of MCT-based structure after Ag ion bombardment and annealing. Inset shows the X-ray diffraction spectrum of the sample in the Bragg configuration. (Reprinted from Smirnov et al. (2017) under the terms of the Creative Commons Attribution License.)

There are two main physical mechanisms to explain preferential sputtering effects in alloy systems – so-called mass effects and so-called bonding effects. In the case of the MCT ternary compound under ion bombardment, the weakness of the Hg-Te bond compared with the Cd-Te bond resulted in the greater ion sputter yield in the Hg-Te system with respect to Cd-Te. The enthalpy of the Hg-Te bond formation is about 0.33 eV and for the Cd-Te bond is about 1.044 eV. It causes Hg, the heaviest element, to have large relative ion sputter yields as compared to other elements and, most likely, plays a dominant role in the preferential sputtering of Te in HgCdTe (Stahle and Helms 1992). Thus, MCT alloy-sputtered surfaces must be depleted in Hg and enhanced in Cd through the weakness of the Hg-Te bond. However, this scenario conflicted with the results of X-ray diffraction analysis of MCT films with the composition 0.23 irradiated with 100 keV silver ions, which proved the formation of polycrystalline MCT layer with x = 0.2 on the surface. According to Sigmund theory (Sigmund 1969), the sputtering yield for MCT is about 11, and effective sputtering depth is about 14 Å. In conformity with our experimental results, the RMS parameter undergoes changes during ion processing in the range of the sputtering depth, and significant erosion is observed after double impact with 140 keV only. Thus, for the MCT ternary compound under ion bombardment, the mass effect is evidently the dominant sputter mechanism, and the preferential sputtering of Cd occurs. In this case, it is possible to assume Cd redeposition followed by oxidation, which is confirmed by the XPS results.

At the same time, knowing that the deformation accumulation is found to lead to the topological instability of the irradiated surface, and being based on our investigations, we can assume that the deformation fields appearing upon implantation of the MCT-based structures are a significant factor of the observed transformation of its surface.

The magnitude of mechanical stresses created in the MCT layer after implantation can be determined from the relation (Cerutti and Ghezzi 1973):

$$\sigma(z) = \frac{C_V(z)\beta E}{(1-\nu)}$$

$$C_V(z) = \left(\frac{\Phi}{\sqrt{2\pi\Delta z}}\right)\exp\left[\frac{(z-\bar{z})^2}{2\Delta z^2}\right]$$

$$C_V(z,\theta) = \left(\frac{\Phi\cos(\theta)}{\sqrt{2\pi\Delta z}}\right)\exp\left[\frac{(z-R_p\cos(\theta))^2}{2\Delta z^2}\right]$$

where:

ν	is Poisson's ratio
E	is the Young modulus
z	is the coordinate
Φ	is a fluence
$C_V(z)$	is the distribution profile of an impurity introduced into the target
$C_V(z, \theta)$	is the distribution profile of an impurity for OI regime of the ion bombardment

The coefficient of MCT crystal lattice contraction by the introduced implant, β, was determined using the results of X-ray diffraction studies of the samples investigated (Smirnov et al. 2013). The mechanical stresses that arise in the near-surface layer of an epitaxial MCT layer attain the maximal value of 1.4×10^3 Pa for boron and $(2.2-2.9)\times10^5$ Pa for silver-implanted structures (see Table 5.1). XRD investigations point to the compression of the boron-implanted MCT and tension of the silver-implanted MCT layers. It was supposed that nano-relief character induced by ion beam irradiation of the MCT depends on the sign of the surface layer deformation.

It was found that the prototype of the MCT detector based on the $Ag_2O/Hg_{0.777}Cd_{0.223}Te/Zn_{0.04}Cd_{0.96}Te$ composite structure exhibited photoresponse in NWIR (0.8–1 μm) as well as MWIR (4–6 μm) and MM spectral range. Preliminary results show that after bombardment of the $Hg_{0.78}Cd_{0.22}Te/Cd_{0.96}Zn_{0.04}Te$ heterostructure with an oblique incident ion beam (45°), the surface became sensitive to irradiation by MM source (140 GHz) in photovoltaic (S_v) mode. In fact, the sensitivity of the composite structure that combines nanostructured ternary semiconductor compounds (MCT) with metal oxide (Ag_2O) inclusions for sub-THz radiation was detected at 300 K. These measurements were performed using lock-in amplifiers with reference modulation at 190 Hz. The value of the measured signal was about 7 ÷ 15 μV at output power ~7 mW. The input signal was detected with NEP at $\nu\approx140$ GHz reached 4.5×10^{-8} W/Hz$^{1/2}$ (Kryshtab et al. 2016). The result of these investigations indicates that composite which integrates the nanostructured ternary compound (HgCdTe) with metal-oxide (Ag_2O) inclusions produced by ion implantation extends the spectral sensitivity region of the basic semiconductor of IR technology to the MM range. This phenomenon can be useful from the viewpoint of developing a new class of electro-optical facility based on MCT that possesses a necessary combination of optical, electro-physical and photoelectric properties.

REFERENCES

Alguero, M., Ricote, J., Torres, M., Amorín, H., Alberca, A., Iglesias-Freire, O., Nemes, N., Holgado, S., Cerveral, M., Piquerasi, J., Asenjo, A. and García-Hernández, M. 2014. Thin film multiferroic nanocomposites by ion implantation. *ACS Appl Mater Interfaces* 6:1909–1915.

Baragiola, R. A. 2004. Sputtering: Survey of observations and derived principles. *Phil Trans R So. Lond A* 362:29–53.

Bayle, M., Bonafos, C., Benzo, P., Benassayag, G., Pécassou, B., Khomenkova, L., Gourbilleau, F. and Carles, R. 2015. Ag doped silicon nitride composites for embedded plasmonics. *Appl Phys Lett* 107:101907.

Belov A. I., Mikhaylov, A. N., Nikolitchev, D. E., Boryakov, A. V., Sidorin, A. P., Gratchev, A. P., Ershov, A. V. and Tetelbaum, D. I. 2010. Formation and "white" photoluminescence of nanoclusters in SiO_x films implanted with carbon ions. *Semiconductors* 44:1450–1456.

Benzo, P., Bonafos, C., Bayle, M., Carles, R., Cattaneo, L., Farcau, C., ... and Muller, D. 2013. Controlled synthesis of buried delta-layers of Ag nanocrystals for near-field plasmonic effects on free surfaces. *J Appl Phys* 113:193505.

Bernas, H. (ed.). 2010. *Materials Science with Ion Beams, Series: Topics in Applied Physics*. Springer, Berlin, vol. 116.

Bernstein, H. W., Sarveswaran, G. H. and Lieberman, K. M. 2004. Sub-10 nm e-beam lithography using cold development of PMMA. *J Vac Sci Technol B* 22:1711–1716.

Boer, A. E., Brongersma, L. M. and Harry, A. 2001. Localized charge injection in SiO_2 films containing silicon nanocrystals. *Appl Phys Lett* 79:791.

Bradley, R. M. and Harper, J. M. E. 1988. Theory of ripple topography induced by ion bombardment. *J Vac Sci Technol A* 6:2390–2395.

Bradley, R. M. and Shipman, P. D. 2012. A surface layer of altered composition can play a key role in nanoscale pattern formation induced by ion bombardment. *Appl Surf Sci* 258:4161–4170.

Carles R., Farau, C., Bonafos, C., Benassayag, G., Bayle, M., Benzo, P., Groenen, J. and Zwick, A. 2011. Three dimensional design of silver nanoparticle assemblies embedded in dielectrics for Raman spectroscopy enhancement and dark-field imaging. *ACS Nano* 5:8774–8782.

Castro, M., Gago, R., Vázquez, L., Muñoz-García, J. and Cuerno, R. 2012. Stress-induced solid flow drives surface nanopatterning of silicon by ion-beam irradiation. *Phys Rev B* 86:214107.

Cerutti, A. and Ghezzi, C. 1973. X-ray observations of induced dislocations at simple planar structures in silicon. *Phys Stat Solidi (a)* 17:273–245.

Chini, T. K., Sanyal, M. K. and Bhattacharyya, S. R. 2002. Energy-dependent wavelength of the ion-induced nanoscale ripple. *Phys Rev B* 66:153404.

Desnica, U. V., Buljan, M., Desnica-Frankovic, I. D., Dubcek, P., Bernstorff, S., Ivanda, M. and Zorc, H. 2004. Direct ion beam synthesis of II–VI nanocrystals. *Nucl Instrum Methods Phys Res A* 216:407–413.

Dhara, S., Datta, A., Wu, C., Chen, K. and Wang, Y. 2005. Mechanism of nanoblister formation in Ga + self-ion implanted GaN nanowires. *Appl Phys Lett* 8:203119–203121.

Dobrovolsky, V., Sizov, F., Kamenev, Y. and Smirnov, A. 2008. Ambient temperature or moderately cooled semiconductor hot electron bolometer for mm and sub-mm regions. *Opto-Electronics Review* 16:172.

Ebe, H., Tanaka, M. and Miyamoto, Y. 1999. Dependency of pn junction depth on ion species implanted in HgCdTe. *J Electron Mater* 28:854–857.

Facsko, S., Dekorsy, T., Koerdt, C., Trappe, C., Kurz, H., Vogt, A. and Hartnagel, H. L. 1999. Formation of ordered nanoscale semiconductor dots by ion sputtering. *Science* 285:1551–1553.

Fekecs, A., Bernier, M., Morris, D., Chicoine, M., Schiettekatte, F., Charette, P. and Arès, R. 2011. Fabrication of high resistivity cold-implanted InGaAsP photoconductors for efficient pulsed terahertz devices. *Opt Mat Express* 1:1165.

Firestein, K. L., Kvashnin, D. G., Sheveyko, A. N., Sukhorukova, I. V., Kovalskii, A. M., Matveev, A. T., Lebedev, O. I., Sorokin, P. B., Golberg, D. and Shtansky, D. V. 2016. Structural analysis and atomic simulation of Ag/BN nanoparticle hybrids obtained by Ag ion implantation. *Mater Des* 98:167–173.

Flagello, D. G. 2007. The future of optical lithography – Extinction or evolution? *Proceedings of 2007 Lithography Workshop*, Rio Grande, Puerto Rico, Dec. 9–13.

Frost, F., Schindler, A. and Bigl, F. 2000. Roughness evolution of ion sputtered rotating InP surfaces: Pattern formation and scaling laws. *Phys Rev Lett* 85:4116.

Gago, R., Vázquez, L., Cuerno, R. Varela, M., Ballesteros, C. and Albella, J. M. 2001. Production of ordered silicon nanocrystals by low-energy ion sputtering. *Appl Phys Lett* 78:3316–3318.

Garaj, S., Hubbard, W. and Golovchenko, J. A. 2010. Graphene synthesis by ion implantation. *Appl Phys Lett* 97:183103.

Garg, S. K., Datta, D. P., Kumar, M., Kanjilal, D. and Som, T. 2014. 60 keV Ar+-ion induced pattern formation on Si surface: Roles of sputter erosion and atomic redistribution. *Appl Surf Sci* 310:147–153.

Garg, S. K., Cuerno, R., Kanjilal, D. and Som, T. 2016. Anomalous behavior in temporal evolution of ripple wavelength under medium energy Ar+-ion bombardment on Si: A case of initial wavelength selection. *J Appl Phys* 119:225301–225307.

Grove, W. R. 1852. On the electro-chemical polarity of gases. *Phil Trans R Soc Lond* 142:87–101.

Henini, M. 1999. EBL opening up the nano-world. *III-Vs Review* 12:18–23.

Holander-Gleixner, S., Williams, B. L., Robinson, H. G. and Helms, C. R. J. 1997. Modeling of junction formation and drive-in in ion implanted HgCdTe. *J Electron Mater* 26:629–634.

Jiaming Z., Qiangmin, W., Ewing, R. C., Jie, L., Weilin, J. and Weber, W. J. 2008. Self-assembly of well-aligned 3C-SiC ripples by focused ion beam. *Appl Phys Lett* 92:3107.

Kanungo, P. D., Kögler, R., Nguyen-Duc, K., Zakharov, N., Werner, P. and Gösele, U. 2009. Ex situ n and p doping of vertical epitaxial short silicon nanowires by ion implantation. *Nanotechnology* 20(16):165706.

Kahng, A. B., Carballo, J.-A. and Gargini, P. 2015. ITRS 2.0: Top-down system integration. *Chip Design* (Spring issue):16–19. http://www.dropbox.com/sh/zkqbpx961v989wb/AAD_foapoTb7 LYzoY4e8rVV_a?dl=0&preview=1504_08_ITRS+2.0_System+Integration.pdf.

Kepaptsoglou, D., Hardcastle, T. P., Seabourne, C. R., Bangert, U., Zan, R., Amani, J. A., Hofsäss, H., Nicholls, R. J., Brydson, R. M. D., Scott, A. J. and Ramasse, Q. M. 2015. Electronic structure modification of ion implanted graphene: The spectroscopic signatures of p- and n-type doping. *ACS Nano* 9:11398–11407.

Krasheninnikov, A. V. and Nordlund, K. 2010. Ion and electron irradiation-induced effects in nanostructured materials. *J Appl Phys.* 107:071301.

Kryshtab, T., Savkina, R. K., Smirnov, A. B., Kladkevich, M. D. and Samoylov, V. B. 2016. Multiband radiation detector based on HgCdTe heterostructure. *Phys Status Solidi C* 13:639. https://doi.org/10.1002/pssc.201510278.

Makeev, M. A., Cuerno, R. and Barabasi, A. L. 2002. Morphology of ion-sputtered surfaces. *Nuclear Instruments and Methods in Physics Research Section B: Beam Interactions with Materials and Atoms* 197(3–4):185–227.

Mazzoldi, P. and Mattei, G. 2005. Potentialities of ion implantation for the synthesis and modification of metal nanoclusters. *Riv Nuovo Cimento* 28:1–69.

Meldrum A., Boatner, L. A. and White, C. W. 2001. Nanocomposites formed by ion implantation: Recent developments and future opportunities. *Nucl Instrum Methods B* 178:7–16.

Mollard, L., Destefanis, G., Baier, N. N., Rothman, J., Ballet, P., Zanatta, J. P., Tchagaspanian, M., Papon, A.-M., Bourgeois, G., Barnes, J.-P., Pautet, C. and Fougères, P. 2009. Planar p-on-n HgCdTe FPAs by arsenic ion implantation. *J Electron Mater* 38:1805–1813.

Moreno-Barrado A., Castro, M., Gago, R., Vázquez, L., Muñoz-García, J., Redondo-Cubero, A., Galiana, B., Ballesteros, C. and Cuerno, R. 2015. Nonuniversality due to inhomogeneous stress in semiconductor surface nanopatterning by low-energy ion-beam irradiation. *Phys Rev B* 91:155303.

Motta, F. C., Shipman, P. D. and Bradley, R. M. 2012. Highly ordered nano-scale surface ripples produced by ion bombardment of binary compounds. *J Phys D Appl Phys* 45:122001-4.

Nastasi, M. and Mayer, J. 2006. *Ion Implantation and Synthesis of Materials*. Springer, Berlin, 2006.

Norris, S. A. 2012. Stress-induced patterns in ion-irradiated silicon: Model based on anisotropic plastic flow. *Phys Rev B* 86:235405.

Oechsner, H. 1975. Sputtering – A review of some recent experimental and theoretical aspects. *Appl Phys* 8:185–198.

Plantevin, O., Gago, R., Vázquez, L., Biermanns, A. and Metzger, T. H. 2007. In situ X-ray scattering study of self-organized nanodot pattern formation on GaSb (001) by ion beam sputtering. *Appl Phys Lett* 91:113105.

Ramaswamy, V., Haynes, T. E., White, C. W., MoberlyChan, W. J., Roorda, S. and Aziz, M. J. 2005. Synthesis of nearly monodisperse embedded nanoparticles by separating nucleation and growth in ion implantation. *Nano Lett* 5:373–377.

Rimini, E. 1995. *Ion Implantation: Basics to Device Fabrication*. Kluwer Academic, Boston.

Rizza, G. and Ridgway, M. C. 2016a. Synthesis of nanostructures using ion-beams: An overview. In: Wesch, W., Wendler, E. (eds) *Ion Beam Modification of Solids*. Springer Series in Surface Sciences, vol 61. Springer, Cham.

Rizza, G. and Ridgway, M. C. 2016b. Ion-shaping of nanoparticles. In: Wesch, W., Wendler, E. (eds) *Ion Beam Modification of Solids*. Springer Series in Surface Sciences, vol 61. Springer, Cham.

Rogalski, A. 2011. *Infrared Detectors*, 2nd edn. CRC Press/Taylor & Francis Group, Boca Raton, FL.

Romanyuk, A. and Oelhafen, P. 2007. Use of ultrasound for metal cluster engineering in ion implanted silicon oxide. *Appl Phys Lett* 90:013118.

Sigmund, P. 1969. Theory of sputtering. I. Sputtering yield of amorphous and polycrystalline targets. *Phys Rev* 184:183.

Simon, C. J. 1986. Garth electron beam testing of ultra large scale integrated circuits. *Microelectron Eng* 4:121–138.

Singh, U. B., Agarwal, D. C., Khan, S. A., Mohapatra, S., Tripathi, A. and Avasthi, D. K. 2012. A study on the formation of Ag nanoparticles on the surface and catcher by ion beam irradiation of Ag thin films. *J Phys D Appl Phys* 45:445304.

Smirnov, V. K., Kibalov, D. S., Krivelevich, S. A., Lepshin, P. A. and Danilin, A. B. 1999. Wave-ordered structures formed on SOI wafers by reactive ion beams. *Nuclear Instruments and Methods in Physics Research Section B: Beam Interactions with Materials and Atoms* 147(1–4):310–315.

Smirnov, V. K., Kibalov, D. S., Orlov, O. M. and Graboshnikov, V. V. 2003. Technology for nano-periodic doping of a metal-oxide-semiconductor field-effect transistor channel using a self-forming wave-ordered structure. *Nanotechnology* 14:709–715.

Smirnov, A. B., Lytvyn, O. S., Morozhenko, V. A., Savkina, R. K., Smoliy, M. I., Udovytska, R. S. and Sizov, F. F. 2013. Role of mechanical stresses at ion implantation of CdHgTe solid solution. *Ukr J Phys* 58:872–880.

Smirnov, A. B. and Savkina, R. K. 2017. Nanostructuring surfaces of HgCdTe by ion bombardment. In: Fesenko, O., Yatsenko, L. (eds) *Nanophysics, Nanomaterials, Interface Studies, and Applications*. NANO 2016. Springer Proceedings in Physics, vol 195. Springer, Cham.

Smirnov, A. B., Savkina, R. K., Gudymenko, A. I., Kladko, V. P., Sizov, F. F. and Frigeri, C. 2014. Effect of stress on defect transformation in B^+ and Ag^+ implanted HgCdTe/CdZnTe structures. *Acta Physica Polonica A* 125:1175–1178.

Smirnov, A. B., Savkina, R. K., Udovytska, R. S., Gudymenko, A. I., Kladko, V. P. and Korchovyi, A. A. 2017. Ion beam nanostructuring of HgCdTe ternary compound. *Nanoscale Res Lett* 12:320.

Smith, B. W. and Suzuki, K. 2007. *Microlithography: Science and Technology*. CRC Press. ISBN: 100824790243.

Stahle, C. M. and Helms, C. R. 1992. Ion sputter effects on HgTe, CdTe, and HgCdTe. *J Vac Sci Technol A* 10:3239.

Stepanov, A. L., Galyautdinov, M. F., Evlyukhin, A. B., et al. 2013. *Appl Phys A* 111:261.

Su, Q., Wang, F., Cui, B., Kirk, M. A., and Nastasi, M. 2016. Temperature-dependent ion-beam mixing in amorphous SiOC/crystalline Fe composite. *Mater Res Lett* 4:198–203.

Technology and manufacturing day of INTEL. 2017. https://www.intc.com/default. aspx?SectionId=817fbab8-2828-44a2-91a0-f10cb8ac2b03&LanguageId=1&EventId=637d9 59b-e595-4f0e-843e-5ee9af9d6520 .

Thompson, M. W. 1981. Physical mechanisms of sputtering. *Phys Rep* 69:335–371.

Thompson, M. W. 1987. The velocity distribution of sputtered atoms. *Nucl Instrum Meth Phys Res B* 18:411–429.

Thompson, M. W. 2002. Atomic collision cascades in solids. *Vacuum* 66:99–114.

Tiwari, S., Rana, F., Hanafi, H., Hartstein, A., Crabbe, E. F. and Chan, K. 1995. A silicon nanocrystals based memory. *Appl Phys Lett* 68:1377.

Torres, M., Ricote, J., Amorin, H., Jaafar, M., Holgado, S., Piqueras, J., Asenjo, A., García-Hernandez, M. and Alguero, M. 2011. *J Phys D Appl Phys* 44:495306.

Trynkiewicz, E., Jany, B. R., Wrana, D. and Krok, F. 2018. Thermally controlled growth of surface nanostructures on ion-modified AIII-BV semiconductor crystals. *Appl Surf Sci* 427:349–356.

Tseng, A. A., Chen, K., Chen, C. D. and Ma, K. J. 2003. Electron beam lithography in nanoscale fabrication: Recent development. *IEEE Trans Electron Packaging Manuf* 26:141–149.

Vieu C., Carcenac, F., Pépin, A., Chen, Y., Mejias, M., Lebib, A., Manin-Ferlazzo, L., Couraud, L. and Launois, H. 2000. Electron beam lithography: Resolution limits and applications. *Appl Surf Sci* 164:111–117.

Viheriälä, J. J., Rytkonen, T., Niemi, T. and Pessa, M. 2007. Narrow linewidth templates for nanoimprint lithography utilizing conformal deposition. *Nanotechnology* 19(1):015302.

Wang, Y., Zhang, D. H., Chen, X. Z., Jin, Y. J., Li, J. H., Liu, C. J., … and Ramam, A. 2012. Bonding and diffusion of nitrogen in the InSbN alloys fabricated by two-step ion implantation. *Appl Phys Lett* 101:021905.

Wei, Q., Zhou, X., Joshi, B., Chen, Y., Li, K.-D., Wei, Q., Sun, K. and Wang, L. 2009. Self-assembly of ordered semiconductor nanoholes by ion beam sputtering. *Adv Mater* 21:2865–2869.

Williams, B. L., Robinson, H. G., Helms, C. R. and Zhu, N. 1997. X-ray rocking curve analysis of ion implanted mercury cadmium telluride. *J Electron Mater* 26:600–605.

Wutzler, R., Rebohle, L., Prucnal, S., Bregolin, F. L., Hübner, R., Voelskow, M., Helm, M. and Skorupa, W. 2015. Liquid phase epitaxy of binary III-V nanocrystals in thin Si layers triggered by ion implantation and flash lamp annealing. *J Appl Phys* 117:175307.

Yastrubchak, O., Domagala, J. Z., Sadowski, J., Kulik, M., Zuk, J., Toth, A. L., Szymczak, R. and Wosinski, T. 2010. Ion-implantation control of ferromagnetism in (Ga, Mn) As epitaxial layers. *J Electron Mater* 39:794–798.

Zhang, R., Wang, Z. S., Zhang, Z. D., Dai, Z. G., Wang, L. L., Li, H., … and Liu, J. R. 2013. Direct graphene synthesis on SiO_2/Si substrate by ion implantation. *Appl Phys Lett* 102:193102.

Ziberi, B., Cornejo, M., Frost, F., and Rauschenbach, B. 2009. Highly ordered nanopatterns on Ge and Si surfaces by ion beam sputtering. *J Phys Condens Matter* 21:224003.

CHAPTER 6

Solid State Composites and Multilayers Produced by Magnetron Sputtering

Larysa Khomenkova and Nadiia Korsunska

CONTENTS

6.1 INTRODUCTION

Solid state composites, being constructed of inorganic/inorganic building blocks, combine the ability of the parent constituents and generate new, sometimes unexpected, properties. Such materials offer a wide functionality that fit the demand of modern microelectronics and photonics. Recent developments of different technological approaches for the synthesis of elemental and complex nanoparticles lend a hand to the creation of numerous inorganic building blocks, allowing the construction of novel materials and structures with unlimited possibilities. The properties of nanocomposite materials depend not only on those of individual building blocks but also on their spatial organisation at different length scales.

Silicon (Si) and germanium (Ge) are the main materials of well-established complementary semiconductor-metal-oxide technology. Optical band gaps of both materials ($E_g^{Si} = 1.12$ eV for Si, and $E_g^{Ge} = 0.66$ eV for Ge) make them very attractive for photovoltaic application. However, being indirect band gap semiconductors, they have very low luminescent efficiency. This fact was the main drawback for the application of both materials in light-emitting diodes.

In 1986, L. Canham reported for the first time on efficient room-temperature infrared photoluminescence (PL) obtained from chemically etched bulk Si surfaces. This emission was very sensitive to chemical pretreatments and gaseous ambient (Canham 1986). After five years, he and, independently, V. Lehmann and U. Gösele, demonstrated the tuning of Si band gap towards the visible spectral range. It was reported that room-temperature PL (Canham 1990) as an absorption edge (Lehmann and Gösele 1991) can be shifted to the red spectral range when bulk Si is transformed to free-standing Si quantum wires by electrochemical etching. Their PL was observed with the naked eye under unfocused green or blue laser line excitation (<0.1 W cm^{-2}). Observed phenomena were attributed to the dramatic two-dimensional quantum confinement (QC) effect in Si nanocrystals (Si-NCs) when their sizes become smaller than the free exciton Bohr radius ($r_B = 4.5$ nm in bulk Si) (Table 6.1). These pioneering works made a strong impact on the reconsideration of silicon as a material whose properties can be tuned significantly by the scaling of physical parameters.

Since 1990, numerous papers were published on various modifications of Si-NCs embedded in different hosts. The publications, appearing mostly in English, are widely available

TABLE 6.1 Material Properties of Silicon and Germanium

Property	Units	Silicon	Germanium
Band gap, E_g, at 300 K	eV	1.12	0.66
Electron affinity, χ	eV	4.0	4.05
Hole mobility, μ_h	cm^{2}V^{-1}s^{-1}	450	1900
Electron mobility, μ_e	cm^{2}V^{-1}s^{-1}	1500	3900
Dielectric constant, $\varepsilon_S/\varepsilon_0$		11.9	16.0
Lattice constant, a	nm	0.543	0.565
Effective mass for electrons, m_e^*/m_0		1.08	0.55
Effective mass for holes, m_h^*/m_0		0.81	0.37
Effective density of states in valence band, N_V, at 300 K	cm^{-3}	1.83×10^{19}	5.65×10^{18}
Effective density of states in conduction band, N_C, at 300 K	cm^{-3}	2.82×10^{19}	1.02×10^{19}
Intrinsic concentration, n_i, at 300 K	cm^{-3}	1.0×10^{10}	2.8×10^{13}
Bohr exciton radius, r_B	nm	4.5	24.3
Thermal conductivity, k, at 300 K	Wcm^{-1}K^{-1}	1.5	0.6
Refractive index, n, at 632.8 nm		3.875	5.441
Melting point, T_m	°C (K)	1412 (1685)	937 (1210)
Density, ρ	g/cm^3	2.33	5.33
Clarke number	%	25.8	6.5×10^{-4}

to the scientific and manufacturing community. The research contributions being presented in different forms as papers, conference proceedings, M.Sc. and Ph.D. thesis, etc., are continuously analysed by experts and are collected in the monographs nearly every two years. Among them, the most addressed are Canham 1997; Ossicini et al. 2003; Pavesi et al. 2003; Siffert and Krimmel 2004; Oda and Ferry 2005; Kumar 2007; Khriachtchev 2009; Koshida 2009; Pavesi and Turan 2010; Torchynska and Vorobiev 2010; Ischenko et al. 2015; Khriachtchev 2016; Chen and Liu 2016; and Cressler 2016.

Such interest in the Si-based nanostructured composite materials was particularly stimulated by the observation of optical gain and electroluminescent emission, the combination of which offers the perspective to create all-in-one-chip, Si-based optoelectronic devices. For this purpose, researchers over the world concentrated their attention on the development of different technological approaches for the fabrication of Si-NCs, monitoring their size distribution and surface chemistry. The understanding of all these aspects is important for the optimisation of the devices and their stable operation. Besides, these efforts brought significant fundamental knowledge in the field of material science and solid-state physics.

It is interesting that *germanium* nanocrystals (Ge-NCs), demonstrating size-dependent optical properties, were discovered by Hayashi et al. in 1982. Although the Ge-NCs have also been addressed in some works (Kanemitsu et al. 1992; Maeda 1995; Min et al. 1996; Chan et al. 2007; Das et al. 2007; Kartopu et al. 2008; Bulijan et al. 2010; Khomenkova et al. 2010), the flow of the publications was significantly lower. The attention of the researchers was particularly renewed in the last five years because Ge-NCs are environmentally friendlier alternatives to classical compound semiconductor nanocrystals, being nontoxic, biocompatible, electrochemically stable and compatible with current microelectronics.

As compared to silicon, germanium is especially appealing due to the higher electron and hole mobility, larger dielectric constant, a larger absorption coefficient and a narrower bulk band gap (Table 6.1), implying the possible tuning of light emission from the near-ultraviolet to the near-infrared. A vital difference between Si and Ge is that Ge has smaller effective masses of electrons and holes and a larger dielectric constant, leading to a significantly larger Bohr exciton radius, r_B (the radius of an electrostatically bound electron-hole pair (exciton) in the bulk material) (Table 6.1). For germanium, the $r_B^{Ge} = 24.3$ nm is much larger than that for silicon ($r_B^{Si} = 4.5$ nm). This fact means that Ge-NCs will impart stronger, more easily identified quantum confinement effects than Si-NCs for the same NC sizes. Moreover, quantum confinement effects will emerge for larger Ge-NCs than for Si-NCs (Kamata 2008). Due to these properties, Ge-NCs are considered to be ideal nodes for memory devices (Chan et al. 2007; Das et al. 2007; Bulijan et al. 2010; Khomenkova et al. 2010; Changa et al. 2011; Haas et al. 2013; Lehninger et al. 2018). Being crystallised at lower temperatures and offering a negative conduction band offset with respect to the Si substrate, the Ge-NCs provide a larger memory window than devices based on Si-NCs with improved retention time. (Khomenkova et al. 2010; Lehninger et al. 2018).

Whether free-standing or embedded in a silicon oxide host, both Si-NCs and Ge-NCs are of great importance largely due to their compatibility with well-established microelectronic technology (Ray 2013; Steimle et al. 2007; Vaughn II and Schaak 2013). These NCs show optical absorption extended over the sun's whole emission spectrum, allowing multiple exciton generation and reduction of energy losses, which is very important for photovoltaic application. The light-to-current conversion efficiency strongly depends on the geometrical dimensions and the crystallinity of the formed NCs, as well as their depth distribution inside the active layer of the photovoltaic cell. The fine-tuning of the mentioned parameters can be achieved via monitoring of fabrication conditions and further materials' post-processing.

In this chapter, the optical, structural and light-emitting properties of the composite materials, based on the different dielectrics (Al_2O_3, HfO_2, ZrO_2 vs SiO_2) with embedded Si-NCs or Ge-NCs fabricated by magnetron sputtering, are considered. Besides, attention is paid to the superlattice approach that offers precise control of size distribution and spatial location of Si-NCs and Ge-NCs. The ways of adjustment of the characteristics of composite materials and superlattices on their basis for specific application are discussed.

6.2 BAND-GAP VARIATION IN Si-NC AND Ge-NC STIMULATED BY QC EFFECT

As was mentioned above, with the decrease of NCs sizes to nanometre regime, when NC diameter (d) becomes the same magnitude as the de Broglie wavelength of the electron wave function, their electronic and optical properties deviate substantially from those of bulk materials. The best description of the carriers inside NCs is as "particles in the box". The reduction of the physical size of this box causes the extension of the energetic levels. In practice, for semiconductor NCs, the expansion of the band gap (E_g) caused by the QC effect follows the relation of $E_g \sim 1/d^a$, where E_g is the band gap and a falls between 1 and 2 (Amato et al. 2014). The variation of the E_g can be extracted from the variation of PL peak position with the NC size that is adequate for free-standing NCs. However, it should be noted that when NCs are embedded in the host material, and/or when their surface-to-volume ratio becomes considerable due to such small NC sizes, this dependence can change to independence of PL peak position on the NC sizes because the PL can originate from surface entities rather than from the NCs themselves.

Figure 6.1 shows this variation of Si-NCs and Ge-NCs. It is seen that this dependence is rather more predictable for Si-NCs than for Ge-NCs. Since Si-NCs show larger exciton

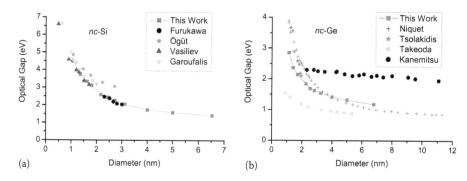

FIGURE 6.1 Comparison of the band-gap energy as a function of NC diameter for silicon (a) and germanium (b). (Reprinted figure with permission from Barbagiovanni et al., Quantum confinement in Si and Ge nanostructures. *J Appl Phys* 111:034307. Copyright 2012, AIP Publishing LLC.)

energy (10 meV for Si against 2.7 meV for Ge), the QC effect is expected to be stronger in Si than in Ge. In this regard, one can assume the easier modulation of Ge-NCs' electronic properties by the QC effect than those of Si-NCs (Cosentino et al. 2011; Amato et al. 2014).

In some cases, the decrease of NC sizes down to the nanometre scale results in uncertainty in the momentum k vectors and, consequently, allows the k selection rules to be broken. This can cause the transformation of indirect band gap towards direct (or quasi-direct), which allows phonon-free electron-hole recombination in the NCs (Barbagiovanni et al. 2012). Besides, the delocalisation of the wave functions in the momentum k space increases overlapping of electron-hole wave functions followed by enhancement of the radiative probability. The shift of the emission wavelength from the infrared to the visible spectral range due to an increase of the band gap can be controlled by the NCs' size. At the same time, a considerable decrease of the number of sites within the NC volume reduces the probability of non-radiative recombination, whereas surface dangling bonds can be controlled by its termination, for instance, by hydrogen or host shells. This approach allows again the monitoring of the radiative recombination and conductivity of the materials.

Room-temperature light emission from Si-NCs and Ge-NCs was widely investigated, and different mechanisms of their luminescence were proposed. However, all of them can be referred to two main types. The first considers that the excitation of carriers and their recombination takes place inside quantum-confined NCs. Another mechanism supposes that the absorption of excitation light occurs inside the NCs, while radiative recombination of carriers happens either at the NC/host interface or via host defects located in the vicinity of this interface. The former mechanism dominates usually in free-standing Si-NCs or Ge-NCs with hydrogen-terminated surfaces, while for the NCs embedded in different dielectric hosts (such as SiO_2, Si_3N_4, SiC, Al_2O_3, HfO_2, ZrO_2, etc.), both mechanisms can be distinguished. The competition of these mechanisms, as well as the manipulation with dangling bond numbers, determines the intensity, spectral peak position and shape of the luminescence band (Delerue et al. 1993; Buuren et al. 1998; Ledoux et al. 2000; Khomenkova et al. 2003; Hessel et al. 2011; Mastronardi et al. 2012; Li et al. 2014; Ni et al. 2014; Khomenkova et al., 2016; Grachev et al. 2017).

It is obvious that the tuning of the NCs' size distribution and spatial location in the host matrix permits the monitoring of optical and electrical properties of composite materials, aiming at their specific applications. Among the different approaches used for Si- and Ge-NCs' production, magnetron sputtering is one of the flexible methods allowing the growing of composite films and superlattices. Hereafter, the properties of Si-NCs and Ge-NCs produced in different host matrices by magnetron sputtering will be considered against fabrication conditions and post-fabrication processing.

6.3 SOLID STATE COMPOSITES AND MULTILAYERS WITH EMBEDDED Si-NCs

6.3.1 Si-NC-SiO$_2$ Materials

During the last decades, silicon oxide (SiO_2) has been the most considered host for Si-NCs. Usually, the formation of Si-NCs in such materials is a two-step process (Figure 6.2). At the first step, the Si-rich-SiO_2 supersaturated solution is produced. The second step supposes

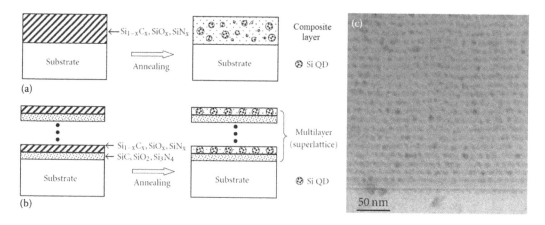

FIGURE 6.2 Schematic illustration of the formation of Si-NCs from phase separation of a single silicon-rich layer (a) and a superlattice (b); (c) TEM images of Si-NCs-SiO$_2$/SiO$_2$ superlattice with distributed Si-NCs. (Reproduced from Cho et al. (2007). Copyright 2007 Hindawi publisher under the Creative Commons Attribution 4.0 International License (http://creativecommons.org/licenses/by/4.0/).)

high-temperature annealing (usually, at 1050°C–1200°C for one hour in nitrogen) caused the formation of Si-NCs via phase separation between Si and SiO$_2$. The size distribution and spatial location of the Si-NCs depends on both excess Si content and annealing conditions (temperature, duration and environment).

The formation of Si-NCs with a uniform mean size and spherical shape is usually required. This demand can hardly be achieved for the composite layers when the possibility of the coalescence of small NCs into large ones exists due to the diffusion process in 3D directions. At the same time, the superlattice approach, where Si-rich sublayers are alternated with SiO$_2$-rich ones, offers a fine-tuning of size distribution and the spatial location of Si-NCs (Zacharias et al. 2002; Gourbilleau et al. 2009b; Khomenkova et al. 2013).

Magnetron sputtering allows easy deposition of Si-rich SiO$_2$ sublayers alternated by SiO$_2$ ones. To achieve the required thickness of each sublayer in the period, the deposition time can be adjusted with a few seconds' accuracy. The periods being stacked is a consequence of the superlattice structure (Figure 6.2). Annealing the superlattice causes the appearance of Si-NCs inside Si-rich layers due to the phase separation process. However, in this case the growing of NCs occurs via lateral Si diffusion, while the SiO$_2$ sublayers suppress such diffusion across the lattice. Sometimes, to achieve the formation of Si-NCs of the required size, a longer annealing time at a higher annealing temperature is required, in comparison with NC formation in a composite layer with the same Si excess. However, the mean size of Si-NCs in a superlattice is more uniform, being determined by the thickness of the Si-rich sublayers. The amount of Si-NCs formed is controlled by the Si excess.

6.3.2 Formation of Si-rich Solid Solution and Express Control of Excess Si Content

The formation of a supersaturated solution by magnetron sputtering can be achieved in several ways: (i) by the sputtering of the composed Si-SiO$_2$ target, which is usually a SiO$_2$ target topped with calibrated Si chips (Koshizaki et al. 1998); (ii) by the sputtering of pure SiO_2 target in mixed argon-hydrogen plasma (Colder at el. 2008; Gourbilleau et al. 2008; Khomenkova et al. 2009); and (iii) by co-sputtering pure Si and the pure SiO_2 target (Khomenkova et al. 2003, 2013).

According to the deposition approach used, the Si content in the films can be tuned via: (i) the variation of the amount of Si chips topped on the SiO_2 target; (ii) the partial pressure of Ar and H_2 gases; and (iii) the RF power density (RFP) applied to the cathode(s). It is worth pointing out that the (ii) and (iii) approaches are more flexible and allow the fabrication of superlattices (Gourbilleau et al. 2009b; Khomenkova et al. 2011, 2013).

The enrichment of the composites in silicon can be proven with optical methods. For instance, *spectroscopic ellipsometry* allows express control of film stoichiometry via the comparison of the refractive index (n) and the absorption coefficient (α) of Si-rich materials with the same parameters of SiO_2 films. In this case, the factor of stoichiometry $x = $ [O]/[Si] for Si-rich films can be extracted from the corresponding refractive index using the method proposed by Dehan et al. (1995). Here, the composite film is considered as nonstoichiometric SiO_x represented by a heterogeneous mixture of amorphous Si and SiO_2 phases to which an effective medium approximation model can be applied. The utility of this approach was shown, for instance, in (Hijazi et al. 2009; Khomenkova et al. 2013). For the films produced by the SiO_2 sputtering in Ar-H_2 plasma, it was shown that the increase of the hydrogen rate ($r_H = F_H/(F_H+F_{Ar})$), where F_H is hydrogen gas flow and F_{Ar} is argon gas flow) allows the variation of the n from $n = 1.487$ ($r_H = 20\%$) up to $n = 1.612$ ($r_H = 80\%$). The Si excess was found to change from 1.5 to 6.7 at %. To achieve the higher Si excess in the films produced with this approach (up to 9 at %), the deposition can be done on heated substrate (Khomenkova et al. 2009; Gourbilleau et al. 2009b).

When Si-rich layers are produced by the co-sputtering of Si and SiO_2 targets, the tuning of Si excess is achieved by the variation of RFP applied to Si cathodes (keeping all other deposition conditions constant). For instance, the increase of RFP_{Si} from 0.74 to 2.96 W/cm² causes the increase of the refractive index from $n = 1.489$ to 1.726 (taken at 1.95 eV light).

The extracted film stoichiometry was found to decrease from $x = 1.722$ to 1.487, reflected by the increase of excess Si content from 0.06 at % up to 10.5 at %. A similar variation of stoichiometry was demonstrated in Khomenkova et al. (2009) and Liang et al. (2012). When higher Si excess should be obtained, further use of higher RFP applied to the Si cathode and lower RFP applied to the SiO_2 target are considered.

The use of certain deposition parameters is determined by the future application of the materials. In principle, both sputtering in Ar-H_2 plasma and co-sputtering allows the production of solid solutions from pure Si via Si-rich-SiO_2 up to pure SiO_2. In the first case, the Si excess is controlled by the hydrogen rate and temperature of the substrate, while in the second phase, it is controlled by RFP applied to Si and SiO_2 cathodes. Sometimes co-sputtering performed in Ar-H_2 plasma also allows finer tuning of Si excess and sublayer thickness via deposition rate (Pratibha Nalini et al. 2011).

Another method allowing an express estimation of excess Si content is *Fourier transform infrared* (FTIR) transmission. The Si content evaluation in this approach is based on the comparison of the peak position of the asymmetric stretching vibration of the oxygen atom in its twofold coordinated bridging bonding site, so-called Si-O TO_3 phonon, for

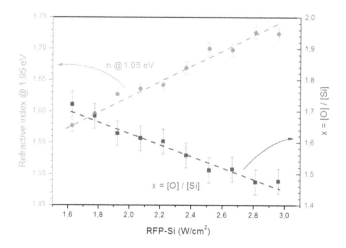

FIGURE 6.3 The refractive index, n (taken at 1.95 eV), and the parameter of layer stoichiometry, $x = [O]/[Si]$, for as-deposited Si-rich SiO_2 layers versus the RFP-Si.

pure SiO_2 and Si-rich SiO_2 materials (Tsu et al. 1989; Kirk 1988; Lange 1989; Innocenzi 2003; Gourbilleau et al. 2009a, 2009c; Hiajazi et al. 2009; Khomenkova et al. 2009, 2013). The continuous random SiO_2 network is known to consist of corner-coordinated SiO_4 tetrahedra, and a disorder of amorphous structures appears from the Si-O-Si bond angle changes. In the mid-infrared range, several Si-O longitude (LO) and transverse (TO) phonons can be detected (Kirk 1988; Lange 1989; Innocenzi 2003). The dominating features of the SiO_2 transmission spectrum observed at normal incidence are the TO_1 Si-O rocking mode (455–460 cm^{-1}), the TO_2 Si-O bending mode (805–810 cm^{-1}), the TO_3 asymmetric Si-O-Si stretch mode with adjacent O-atoms in phase (1050–1085 cm^{-1}) and the TO_4 asymmetric Si-O-Si stretch mode with adjacent O-atoms out of phase (1150–1200 cm^{-1}). At the Brewster angle incidence, the LO phonons of corresponding Si-O modes appear as LO_1 (506–510 cm^{-1}), LO_2 (~820 cm^{-1}), LO_3 (1256–1260 cm^{-1}) and LO_4 (about 1200 cm^{-1}). Among all Si-O related phonons, the LO_3 and TO_3 phonons are the most sensitive to the film stoichiometry. As Schmidt and Schmidt (2003) reported, the TO_3 mode is sensitive to both the bonding configuration and the SiO_x composition, suggesting that it is made up from sub-bands with variable relative weight. This band consists of two sub-bands centred at 1054 and 1090 cm^{-1} corresponding to Si-O-Si bond angles of 131° and 143°, respectively (Lisovskii et al. 1995), the 1050 cm^{-1} band having an increased weight in SiO_2 films of higher density. In the case of SiO_x ($x < 2$), the TO_3 peak is made up from four major sub-bands (1000, 1038, 1066 and 1092 cm^{-1}) corresponding to the several (Si-O_{4-k}-Si_k, $0 \leq k \leq 4$) bonding tetrahedra to which the Si next neighbours of the vibrating O-atom may belong. The band at 1092 cm^{-1} is assigned to the Si-(O_4) tetrahedron, dominating the spectrum in the case of thermally grown or annealed SiO_2 films. The formation of Si-NCs during thermal annealing of SiO_x leads to a higher fraction of Si-O-Si configurations in which the two Si next neighbours of the vibrating O-atom are part of either Si-(O_4) or Si-(O_3Si) bonding tetrahedra. As a consequence, the relative weight of the 1092 cm^{-1} sub-band increases upon SiO_x annealing.

The LO_3 phonon is a feature of the Si-O-Si bond with the angle of 180° that is evidence of the formation of an Si/SiO_2 interface of a good quality (Ono et al. 1998; Gourbilleau et al. 2008; Khomenkova et al. 2012). Thus, the increase of the weight of the LO_3 mode in the FTIR spectra of annealed samples gives evidence on the formation of an Si/SiO_2 interface of good quality when Si nuclei crystallise. Thus, the specific modification of FTIR spectra can be used to quantify the progress in phase separation and formation of Si-NCs achieved with a particular annealing regime.

It should be noted that for correct estimation of film stoichiometry from FTIR data, the comparison of the spectra measured for Si-rich-SiO_2 and pure SiO_2 films of the same thickness is required. This demand is caused by the effect of SiO_2 layer thickness on the peak position of LO_3 and TO_3 phonons, i.e. the decrease of SiO_2 layer thickness to nanometre scale is accompanied by a shift of the peak positions of LO_3 and TO_3 phonons (about 25–30 cm⁻¹) towards the lower wavenumbers (Ono et al. 1998). It is followed also by the decrease of the contribution of LO_4-TO_4 phonon doublet that is usually considered to be responsible for SiO_2 disordering. It is obvious that longer sputtering in argon plasma results in the thickening of the film and doesn't affect the film stoichiometry. However, the transformation of FTIR spectra with the increase of film thickness is significant (Figure 6.4a). This can be caused by the rearrangement of Si-O bonds via an adjustment of Si-O bond angles rather than by the variation of excess Si content with deposition time, especially for SiO_2 films.

The approach described above was already used for the evaluation of Si excess in Si-rich SiO_2 composites (Khomenkova et al. 2012) and showed more precise results in comparison with those reported elsewhere (Ternon et al. 2002).

Figure 6.4b demonstrates the FTIR spectra of as-deposited Si-rich SiO_2 layers grown with different RFP_{Si}. The intense absorption band around 1000 cm⁻¹ corresponds to Si-O TO_3 phonon (Figure 6.4b). When RFP_{Si} increases, the intensity of this phonon decreases, accompanied by its broadening and shift towards the lower wavenumbers. Along with this, a significant decrease of the intensity of LO_3 phonon, followed by the increase of the LO_4-TO_4 doublet contribution, is observed (Figure 6.4b). Since the thickness of the different layer is about 56–60 nm, this evolution of the FTIR spectra can be explained by the distortion of Si-O bonds due to pronounced Si incorporation into the SiO_2 host.

Film stoichiometry was estimated based on the analysis of FTIR spectra shown in Figure 6.4b and c. As one can see from Figure 6.4c, the TO_3 phonon of pure SiO_2 is centred at $v_{TO_3} = 1060$ cm⁻¹. This value is lower than the TO_3 phonon peak position referenced often for thermal SiO_2 ($v_{TO_3} = 1080$ cm⁻¹) (Kirk 1988; Lange 1989; Innocenzi 2003). This difference in peak positions can be due to lesser thicknesses of our SiO_2 layer as well as by its microstructure caused by the angle of Si-O-Si bond variation (Tucker et al. 2005). As was reported by Pai et al. (1986), TO_3 phonon peak position (v_{TO_3}) depends linearly on film stoichiometry x. It was proposed to use the relation $x = 0.02 \cdot v_{TO_3} - 19.2$ to extract the x values. Applying this formula for our samples (Figure 6.4c), a linear decrease of the x from 1.67 to 1.49 was obtained that corresponds to the increase of Si excess from 6.4% to 10.4% when RFP_{Si} rises 1.63 to 2.96 W/cm². One can see good agreement by comparing these results

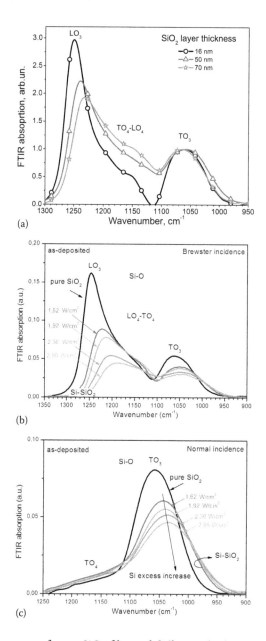

FIGURE 6.4 (a) FTIR spectra of pure SiO_2 films of different thickness achieved via adjusting of deposition time. (b, c) FTIR spectra of Si-rich SiO_2 films produced at RFP_{Si} increased from 1.63 to 2.96 W/cm^2 measured at Brewster (b) and normal (c) incidence of excited light. The spectra for the pure SiO_2 layer grown with the same conditions from the SiO_2 target only is also presented for comparison. The thickness of all samples is about 60 nm.

with those obtained from ellipsometry data. The Si excess estimated for the films grown with Ar-H_2 plasma was found in Liang et al. (2012a) to have similar dependence.

It is worth pointing out that some differences between the stoichiometry factor extracted from ellipsometry and FTIR data can be found for the case of high Si content. The Si excess extracted from FTIR spectra could be a bit lower than that found in ellipsometry data.

This effect is artificial and caused by the FTIR insensitivity to the Si-Si vibration modes inside of some Si agglomerates that can be formed in the Si-rich samples with high Si content. On the other hand, such difference allows expecting the formation of amorphous Si nuclei or agglomerates in as-deposited Si-rich-SiO$_2$ films. Such effects were reported in Khomenkova et al. (2016) and will be discussed in Section 6.3.3.

6.3.3 Effect of Thermal Treatment on the Formation of Si-NCs

Let us consider *the effect of annealing treatment* on the transformation of film microstructure and Si-NC formation. The information about these processes can also be extracted from FTIR spectra. Their main transformation occurs via the shift of Si-O TO$_3$ and LO$_3$ phonons towards the side of higher wavenumbers. As one can see from Figure 6.5, an increase of annealing temperature (T_A) from 600 to 1150°C results in the gradual shift of both phonons as well as their narrowing. Specifically, the maximum TO$_3$ phonon shifts from about 1047 cm^{-1} to 1076 cm^{-1} (Figure 6.5a, c), while the peak position of LO$_3$ phonons is moving from about 1200 cm^{-1} to 1253 cm^{-1} (Figure 6.5b, c). Finally, LO$_3$ and TO$_3$ bands reach the spectral positions that are apparently close to those of pure SiO$_2$ (1076 cm^{-1} and 1255 cm^{-1}). Along with this, the intensity of LO$_3$ phonon increases significantly, whereas the contribution of LO$_4$-TO$_4$ doublet decreases (Figure 6.5a, b).

Figure 6.5c shows the variation of the ratio of the intensities of LO$_3$ and TO$_3$ phonons, I_{LO3}/I_{TO3}, versus annealing temperature. As one can see for $T_A \leq 700$°C, the ratio I_{LO3}/I_{TO3} is nearly constant, whereas the v_{TO_3} and v_{LO_3} shifts from 1047 cm^{-1} to 1065 cm^{-1} and from 1200 cm^{-1} to 1225 cm^{-1}, respectively (Figure 6.5). These facts can be explained by the reconstruction of the Si-O and Si-Si bonds in Si-SiO$_{4-x}$ and Si-Si$_{4-x}$ tetrahedral and by the initiation of a

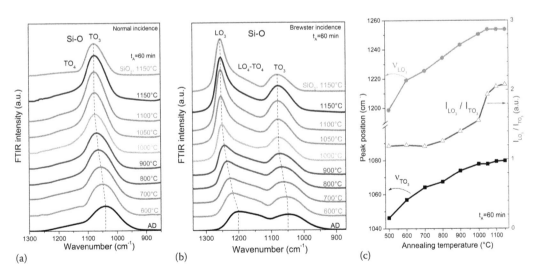

FIGURE 6.5 Evolution of FTIR spectra of Si-rich SiO$_2$ composite film, recorded with normal (a) and Brewster (b) incidence of excited light. Spectra are shifted in vertical direction for clarity. (c) Variation of the peak positions of TO$_3$ (v_{TO_3}) and LO$_3$ (v_{LO_3}) phonons and the ratio of their intensities with T_A. Annealing time for all the figures is $t_A = 60$ min. The Si-rich SiO$_2$ film was grown with RFP$_{Si} = 2.36$ W/cm^2.

phase separation process. When $T_A = 800°C–1000°C$, the ratio I_{LO_3}/I_{TO_3} increases gradually, whereas both phonons continue their shift to the higher wavenumbers up to $v_{TO_3} = 1076$ cm^{-1} and $v_{LO_3} = 1253$ cm^{-1}. This gives evidence of the formation of Si nuclei and small clusters being more prominent at higher T_A. For $T_A > 1000°C$, the variation of v_{TO_3} and v_{LO_3} is negligible and similar to those of pure SiO$_2$ film (Figure 6.5a, b). However, the ratio I_{LO_3}/I_{TO_3} increases sharply from $T_A = 1000°C$ to $T_A = 1050°C$, being afterwards unchangeable (Figure 6.5c). As was mentioned above, a high intensity of narrow Si-O LO$_3$ phonon is usually considered as an evidence of the formation of perfect Si/SiO$_2$ interface (Olsen et al. 1989). The behaviour of the ratio I_{LO_3}/I_{TO_3} and the peak positions of LO$_3$ and TO$_3$ phonons can be considered as an indication of the crystallisation of Si nuclei and the appearance of Si-NCs embedded in the SiO$_2$ host. The low contribution of LO$_4$-TO$_4$ doublet testifies to the significant improvement of matrix structure. Thus, the results described above demonstrate the utility of the FTIR method for the evaluation of the properties of composite films versus deposition conditions and post-deposition processing. Higher temperature treatment stimulates microstructure transformation that modifies optical, electrical and luminescent properties.

6.3.4 Structural Properties of the Composites and Mechanism of Si-NC Formation

To observe the Si-NCs and estimate their sizes and distribution, several structural approaches are usually used: X-ray diffraction and Raman scattering spectroscopy, as well as high-resolution (HR) or energy-filtered (EF) transmission electron microscopy (TEM). The first two methods are non-destructive and provide information based on the comparison of the data obtained for the composites with Si-NCs with those of bulk Si.

The typical Raman scattering spectra and XRD pattern for the films with Si-NCs are shown in Figure 6.6. In the case of Raman scattering, the peak position and full-width of Si TO phonon of bulk Si and those of Si-NCs-SiO$_2$ composites are compared

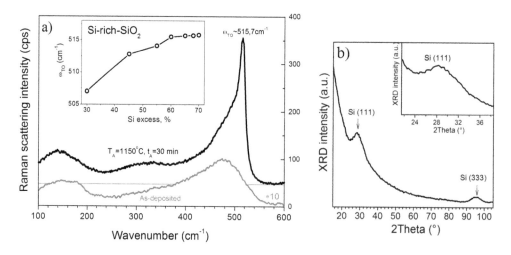

FIGURE 6.6 Typical Raman spectrum of Si-rich-SiO$_2$ layer grown on fused silica substrate before and after annealing at 1150°C for 30 min in nitrogen. Insert shows the variation of TO phonon peak position with Si excess in annealed samples.

(Figure 6.6a). The broadening of the spectrum and its shift to the lower wavenumber side in comparison with that of bulk Si testifies to the formation of Si-NCs.

The mean size of Si-NCs was estimated by the fitting of Raman spectra, applying the procedure described by Fauchet and Campbell (1988) and developed in Doğan and van de Sanden (2013). Thus, the Si-NCs' mean size was found to be changed from about 2.3 nm up to 6.1 nm. The formation of Si-NCs was also confirmed by the appearance of (111) and (333) reflections for crystalline Si in the corresponding XRD pattern (Figure 6.6b). Using the Debye-Scherrer formula for the (111) reflections the mean size of Si-NCs was also extracted and found to be similar to the values mentioned above (Khomenkova et al. 2016).

It is worth pointing out that for fine determination of Si-NCs' size distribution and, what's more important, spatial location of Si-NCs in the host matrix, these composites were studied with the TEM and HR-TEM methods. However, amorphous Si nanoclusters (Si-NCs) or small Si-NCs (less than 2 nm in diameter) are hardly observed with HR-TEM due to the small amount of diffracted crystalline planes. In such cases, the EF-TEM method is used, being sensitive to both crystallised and amorphous Si inclusions and allowing accurate size distribution measurement (Schamm et al. 2008; Pratibha Nalini et al. 2012). Furthermore, Si phase identification depends on the deconvolution of Si peaks on electron energy loss spectroscopy (EELS) spectra and on contrast enhancement. Typically, this can lead to some changes in the Si-NCs' size distribution depending on the data treatment applied to the TEM images. Besides, it gives access only to planar projections that suffer from a superposition of several particles, in the case of a multilayered structure.

Usually Si-NCs grown in composites with low excess Si content have spherical shapes with abrupt interfaces (Figure 6.7a). At the same time, the Si-NCs embedded in the composites with high Si excess show non-spherical shapes (Figure 6.7b). As one can see, the Si-NCs are elongated in (111) direction, which is the preferable direction of Si lattice construction (Figure 6.7).

As it was mentioned above, the Si-NCs' formation in supersaturated SiO_x material occurs via phase separation stimulated by annealing treatment. This process depends on

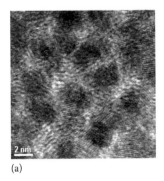
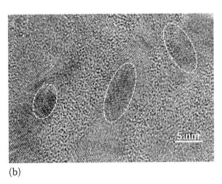

(a)   (b)

FIGURE 6.7   HR-TEM images obtained for the Si-NCs-SiO_2 films with low (below 15%) (a) and high (about 50%) (b) excess Si content. The formation of spherical Si-NCs can be distinguished in (a), whereas for high Si content, elongated Si-NCs are formed.

the excess Si content. Earlier reports show that Si-NCs form via the appearance of Si nuclei and their Ostwald ripening governed by Si diffusion (Hubert 1980; Nesbit 1985; Rochet et al. 1988; Barranco et al. 2001). In other publications it is considered that Si nuclei appear due to spinodal decomposition of nonstoichiometric SiO_x, whereas at high temperatures (more than 1000°C), Ostwald ripening can occur and results in the growing of large Si-NCs due to the diffusion of Si from smaller ones. It is considered that for the Si-rich SiO_2 composites with Si excess below 10 at %, the nucleation process dominates, while for high Si excess content, spinodal decomposition is responsible for the formation of Si nuclei and their crystallisation, followed by the slow growing of already formed Si-NCs due to the dissolving of small Si-NCs.

Taking into account the mechanisms described above, as well as considering experimental data obtained by FTIR, Raman scattering and TEM methods, one can analyse the effect of annealing treatment on the formation of Si-NCs in our samples. In the materials described above, two main mechanisms are possible. For the samples with low Si excess (below 10 at %), one can expect that the main role is played by the nucleation of Si seeds and their growing in size. In these materials, the diffusion of silicon is the main factor that controls Si-NCs' formation. The growing of Si inclusions via Ostwald ripening and their crystallisation requires both higher temperatures and a longer annealing time.

In supersaturated Si-rich composites with a higher Si excess, an appearance of Si nuclei occurs via spinodal decomposition and governed by oxygen out-diffusion from some spaces towards interface shells between Si seeds and SiO_2 hosts. The latter requires a shorter annealing treatment due to the high diffusion coefficient of oxygen. Then the crystallisation of already formed Si seeds occurs, while Ostwald ripening cannot be ruled out. It is worth pointing out that such decomposition requires a shorter annealing time and allows Si-NCs' formation via rapid thermal treatment.

The materials described above were used to produce Si-NCs-SiO_2/SiO_2 superlattices. However, before starting their consideration, one can address the luminescence from corresponding Si-rich layers.

6.3.5 Luminescence of Si Nanocrystals in SiO_2 Films

Luminescence of Si-NCs-SiO_2 composites. In most cases, as-deposited samples did not show any PL emission. However, for the samples deposited on heated substrate (at 500°C–600°C), a weak red PL band centred at about 730 nm was detected (Figure 6.8a, b). It was assumed to be caused either by some SiO_x defects or by some Si agglomerates formed due to high substrate temperatures. Annealing treatment at higher temperatures stimulates an enhancement of visible emissions. Figure 6.8c represents the PL spectra of the films grown with different RFP_{Si} values and annealed at 1100°C. As one can see, the higher the excess Si content in the layers, the more pronounced is the shift of PL peak position to near-infrared spectral range that can be observed. For instance, PL peak position shifts from about 745 nm ($x = 1.72$) to about 845 nm ($x = 1.49$), followed by the broadening of PL spectrum. The analysis of these spectra shows that long-wavelength shifts of the PL

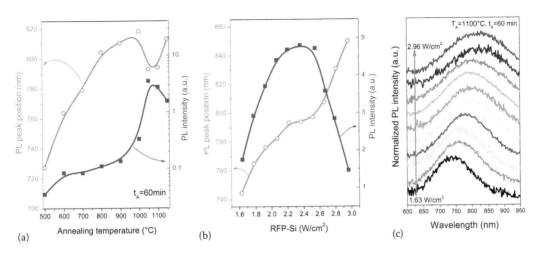

FIGURE 6.8 (a) Variation of PL intensity and peak position with annealing temperature for the film with $x = 1.46$ (grown at $RFP_{Si} = 2.36$ W/cm^2). The data at $T_A = 500°C$ corresponds to the as-deposited samples (fabricated at 500°C); (b) Effect of RFP_{Si} on the PL peak position and intensity. (c) PL spectra of the composites grown with different RFP_{Si} and annealed at 1100°C for 60 min (a).

peak position with Si content rise (Figure 6.8b) that can be explained by the formation of Si-NCs with a larger mean size and wider size distribution. The non-monotonous changing in intensity is assumed to be caused by the increase of the number of Si-NCs when RFP_{Si} increases up to 2.2 W/cm^2. However, further rise of Si content ($RFP_{Si} > 2.5$ W/cm^2) results in the decrease of PL intensity that can be due to growth in the sizes of Si-NCs and the increasing role of the non-radiative channel (for example, Si dangling bonds due to lack of oxygen for saturation).

The effect of annealing temperature on the PL properties of the films grown with $RFP_{Si} = 2.36$ W/cm^2 is demonstrated in detail in Figure 6.8a. The shift of PL peak position to the longer wavelength side from 730 nm ($T_A = 500°C$) to 820 nm ($T_A = 1000°C$) is seen (Figure 6.8a). Taking into account the FTIR data for the same samples (Figure 6.5), this PL behaviour can be explained by the formation of amorphous Si nuclei (or agglomerates) as well as their growing in the sizes and an increase of their number with T_A rise. When $T_A = 1000°C–1050°C$, a sharp increase of PL intensity and the "blue" shift of PL peak position are observed (Figure 6.8). This can be caused by the crystallisation of Si agglomerates and the formation of Si-NCs accompanied by Si/SiO$_2$ interface shell appearance. The latter provides the better confinement of the carriers inside Si-NCs leading to higher PL emission. Further T_A rise up to 1150°C results in a PL quenching and a slight shift of the PL peak position to 810 nm (Figure 6.8). This shift can be due to the expansion of the Si-NCs via a coalescence of the smaller Si-NCs, whereas the PL quenching can be explained by the decrease of the Si-NCs' number as well as a significant contribution of non-radiative channels (e.g. Si dangling bonds). The highest PL intensity was observed for the films with $x = 1.5–1.6$ (grown with $RFP_{Si} = 2.2–2.5$ W/cm^2). These Si-rich materials were chosen for the production of Si-rich-SiO$_2$/SiO$_2$ superlattices.

PL emission of Si-rich-SiO₂/SiO₂ superlattices. Taking into account the results described above for single layers, we fabricated superlattices with some optimal deposition conditions. Thus, we used $RFP_{SiO2} = 8.88$ W/cm² and $RFP_{Si} = 2.36$ W/cm² ($x = 1.5$). The thickness of Si-rich-SiO₂ and SiO₂ sublayers was fixed at 3 nm. This choice was made based on our previous results (Khomenkova et al. 2013) and taking into account those reported by other authors (Zacharias et al. 2002; Gourbilleau et al. 2009b; Pratibha Nalini et al. 2011). It was shown that thinner SiO₂ layers cannot efficiently confine the Si atoms in Si-rich layers, and upon annealing, Si out-diffusion from Si-rich layers is significant. Thicker SiO₂ layers (more than 3.8 nm) favour the physical confinement of Si atoms inside Si-rich sublayers, and the better confinement of the carriers in the formed Si-NCs stimulates an enhancement of PL intensity. However, in terms of the future application of such superlattices, the thicker SiO₂ sublayers increase the resistivity of the structures. Thus, hereafter the properties of the superlattices with 3-nm-SRSO and 3-nm SiO₂ sublayers will be described.

The PL spectra of Si-rich-SiO₂/SiO₂ superlattices are shown in Figure 6.9. As one can see, the increase of annealing temperature from 600°C to 900°C doesn't affect the PL peak position observed at about 760 nm, whereas it results in the increase of PL intensity. Further T_A rise stimulates the PL enhancement, reaching some saturation for $T_A = 1100°C–1150°C$ (Figure 6.9). The comparison of this PL behaviour with that observed for a single Si-rich layer (Figure 6.8) shows that the formation of Si-NCs in the superlattices requires higher annealing temperatures due to 2D-confinement of Si diffusion. Similar results were observed in Zheng and Li (2005) and explained by the appearance of strain field at Si-rich-SiO₂/SiO₂ interfaces confining lateral diffusion of Si atoms in Si-rich sublayers. This allows supposing

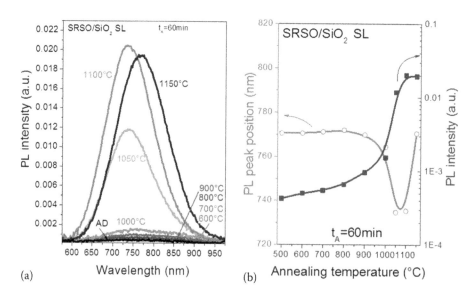

FIGURE 6.9 (a) Evolution of PL emission of Si-rich-SiO₂/SiO₂ superlattices with annealing temperature, T_A, mentioned in the figure. $t_A = 60$ min. (b) analysis of the PL spectra versus annealing temperature.

an amorphous nature of the Si agglomerates after the annealing of the superlattices at $T_A \leq 900°C$, whereas annealing at 1100°C–1150°C favours Si agglomerates crystallisation.

Structural properties of the Si-rich-SiO$_2$ composites and superlattices. To support our findings from optical characterisation of the samples, they were submitted to the investigation with the transmission electron microscopy method. Transmission electron microscopy (TEM) and energy-filtered TEM (EF-TEM) observations were performed. The EF-TEM approach allows the presence of the Si-NCs, whatever their orientation towards the electron beam, to be evidenced when the image is filtered with 17 eV energy corresponding to Si plasmon peak. Besides, EF-TEM allows also the non-crystallised Si agglomerates and inclusions to be revealed. Figure 6.10a and b shows the EF-TEM results obtained for the Si-rich-SiO$_2$ composite annealed at 900°C for 60 min. As one can see from EF-TEM images filtered at 0 eV (a) and 17 eV (b), there are numerous amorphous Si agglomerates in the sample volume. However, only a small amount of them could be found as Si-NCs, as shown by the HR-TEM image in Figure 6.10c. Since for the samples annealed at $T_A < 900°C$, the

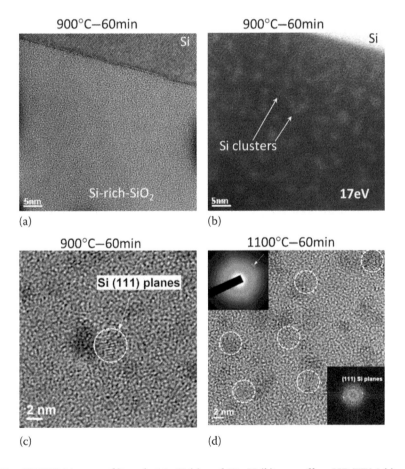

FIGURE 6.10 EF-TEM images filtered at 0 eV (a) and 17 eV (b), as well as HR-TEM (c) of the composite Si-rich-SiO$_2$ sample annealed at $T_A = 900°C$. (d) HR-TEM image of the sample annealed at 1100°C. For all the figures, the RFP$_{Si} = 2.36$ W/cm^2, and t$_A = 60$ min.

formation of Si-NCs was not revealed by any method, one can suppose that $T_A = 900°C$ is the "set-in" temperature when the crystallisation of Si agglomerates starts.

The sample annealed at 1100°C shows the formation of crystallised Si-NCs with a mean diameter of about 5 nm (Figure 6.10d). The selective diffraction pattern testifies to their formation by the defined ring, corresponding to the diffraction from the plane (111) of Si. Thus, such results confirm our conclusion drawn from the analysis of optical properties of the samples about significant crystallisation of Si-NCs at $T_A > 1000°C$. EF-TEM study revealed the formation of amorphous Si agglomerates in the films annealed at 700°C. This can happen due to phase separation at $T_A = 700°C–900°C$, while at $T_A = 900°C$, the crystallisation of some Si agglomerates can be revealed.

The comparison of the TEM data obtained for composite layers and superlattices revealed the similarity in the decomposition process in studied materials. The agglomeration of silicon entities appears at $T_A = 700°C–900°C$, but they keep their amorphous nature (Figure 6.11a, b), while pronounced crystallisation was observed for $T_A = 1100°C–1150°C$ (Figure 6.11c, d).

It is worth pointing out that the effect of thermal treatment on the properties of Si-rich-SiO_2 composites and superlattices described above is very similar to that reported for Si-rich-SiO_2 materials produced with reactive argon-hydrogen sputtering (Gourbilleau et al. 2009a, 2009c; Pratibha Nalini et al. 2012). Some TEM and EF-TEM results for such materials are collected in Figure 6.12.

It is seen that bright-filed TEM (Figure 6.12a) confirms the formation of a multilayer structure. The EF-TEM data showed the presence of Si-NCs inside Si-rich sublayers being well-distributed along it (Figure 6.12b). The plan view image (Figure 6.12c) evidences the presence of a high density of Si-NCs (about 10^{19} cm^{-3}). Their mean size is about 3 nm (inset in Figure 6.12b), which fits the thickness of Si-rich sublayers well. The average distance between Si-NCs was estimated to be about 1.5 nm (Figure 6.10b).

Such similarity in the properties of Si-rich SiO_2 materials could originate from the similar Si excess content that can be the main factor responsible for materials properties. The analysis of the samples produced by different approaches but having similar Si excess

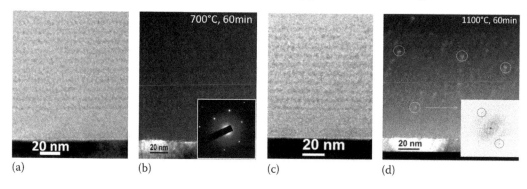

(a)   (b)   (c)   (d)

FIGURE 6.11 Bright (a, c) and dark field (b, d) TEM images of Si-rich SiO_2/SiO_2 superlattice annealed at $T_A = 700$ C (a, b) and 1100 C (c, d) for 60 min in nitrogen flow. The inset in (b) is a selected area electron diffraction (SAED) image confirming the amorphous nature of the sample. The inset in (d) is a filtered Fourier transform (FFT) image from a bright spot (marked as a ring) showing the (111) Si planes, and thus, the formation of crystallised Si-NCs.

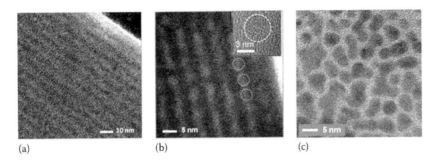

(a)   (b)   (c)

FIGURE 6.12   Bright field TEM (a), EF-TEM of superlattice cross section (b) and plan view (c) of Si-rich-SiO_2/SiO_2 superlattices with Si-NCs. Inset in (b) shows HR-TEM image of a Si nanocrystal. (Reproduced with permission from Gourbilleau et al., *J Appl Phys*, 106, 013501, 2009.)

showed the similarity of their structural and optical properties as it was reported in the references cited in this session. It should be pointed out that such significant attention paid to Si-NCs-SiO_2 composites in the present chapter is caused by their application in optoelectronics and photovoltaics.

6.3.6 The Future of Si-NCs-SiO_2 Materials

The properties of Si-NCs-SiO_2 composites and superlattices have been deeply investigated. Monitoring Si-NCs' distribution and spatial localisation allows tuning their electrical and optical properties. Different applications of Si-NCs-SiO_2 composites in microelectronics, photovoltaics and optoelectronics were announced (Khriachtchev 2016). However, Si-NCs-SiO_2 materials possess high ohmic resistance. To improve their electrical properties, the application of Si-NCs-SiO_2/SiO_2 superlattices with a high density of Si-NCs in Si-rich-SiO_2 sublayers and thin SiO_2 sublayers was proposed (Ding et al. 2011; Roussel et al. 2011; Khomenkova et al. 2013). Further optimisation of such structures was achieved via the substitution of SiO_2 sublayers by Si_3N_4 ones (Pratibha Nalini et al. 2011; Cho et al. 2007).

As described above, SiO_2 is still the best matrix for Si-NCs. In this regard, the modification of the fabrication approach is still considered to be an option for the improvement of superlattice properties. Recently, Fu et al. (2016) showed that application of hydrogen ion beam treatment (HIB) of Si-NCs-SiO_2 superlattices allows one to enhance the efficiency of electroluminescence with reduced operating current density.

Figure 6.13 shows the HR-TEM cross section images of the 10-period superlattices with clear interfaces between Si-rich SiO_2 and SiO_2 sublayers. The estimated thickness of each sublayer in the period is about 1.5 nm. It is seen that no crystalline Si-NCs were observed in the non-irradiated sample (Figure 6.14a). The fabricating of Si-NCs within a Si-rich-SiO_2 layer with a thickness of less than 1.6 nm was difficult because an extremely high Gibbs free energy was introduced during the formation of Si-NCs by increasing the amount of amorphous/crystalline Si and oxide/crystalline Si interfaces. In addition, the thickness of SiO_2 was less than 2 nm, and therefore, the nucleation of the Si-NCs was difficult. It was because the appearance of a rough SiO_2/Si-rich-SiO_2 interface reduced the driving forces

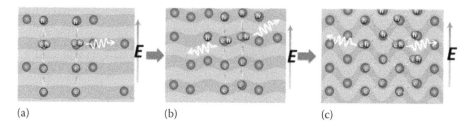

(a) (b) (c)

FIGURE 6.13 Schematic of the variation of the interface between Si-NCs-SiO$_2$ (blue) and SiO$_2$ (grey) sublayers induced by employing a HIB treatment with the energies of 0 (a), 70 (b) and 116 eV (c). The blue dots represent the Si-NCs. As the HIB energy increases, the number of Si-NCs increases and nanoscale interface becomes nano-roughened.

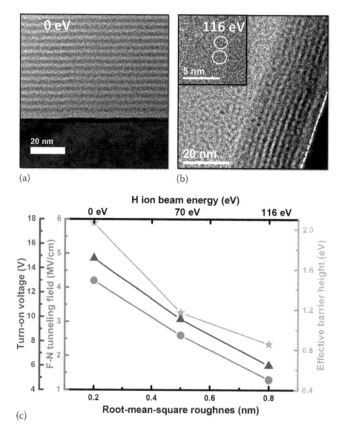

FIGURE 6.14 HR-TEM micrographs of Si-NCs-SiO$_2$/SiO$_2$-superlattice structures annealed at 1000°C for 3 hours in mixed gas (N$_2$ (95%) and H$_2$ (5%)) without (a) and with (b) HIB (116 eV) treatment. The inset in (b) illustrates the formation of Si-NCs with size less than 2 nm in the irradiated sample. (c) The threshold Fowler–Nordheim (F-N) tunnelling electric field, turn-on voltage and effective barrier height as a function of the HIB energy and root-mean-square roughness.

of the self-assembled Si-NCs formation. Despite these limitations, crystalline Si-NCs were obtained in the sample after HIB treatment with 116 eV, as shown in the inset of Figure 6.14b. The size of the Si-NCs was approximately 1.6 nm.

Another perspective approach is to use free-standing Si-NCs embedded in a polymer host. The Si-NCs are produced with a chemical route that allows achieving uniform size distribution. Such Si-NCs, being covered with some shell or terminated by hydrogen, find their application in smart windows, biological labelling and photovoltaic and light-emitting devices.

6.4 PROPERTIES OF ALTERNATIVE COMPOSITES WITH Si-NCs

To overcome SiO_2 bottleneck, described above, it was proposed to use either silicon (oxy) nitride or (oxy)carbide as an alternative host matrix for Si-NCs (Di et al. 2010; Pratibha Nalini 2011; Sohrabi et al. 2013). The Si-NCs formed in Si_3N_4 layers show a low barrier height, the highest density and less silicon requirement during deposition than those in SiO_2. Besides, Si-NCs-Si_3N_4 are preferable because they require a lower thermal budget (800°C–850°C) in comparison to Si-NCs formed in Si (oxy)carbides (about 1100°C) (Sohrabi et al. 2013).

Providing the highest quality of electrical isolation as gate material, SiO_2 is faced with the problem that it does not respond anymore on the demand of continuous aggressive downscaling of microelectronic devices. Nowadays, alternative dielectrics such as HfO_2, $HfAl_xO_y$, Al_2O_3, ZrO_2 and Si_3N_4 have become more attractive. Owing to their higher permittivity ($k \geq 7$ contrary to $k = 3.9$ for SiO_2), they allow the achievement of advanced properties of memory structures with embedded Si-NCs (Ilse et al. 2016). However, such materials are rarely addressed in optical applications. At the same time, they have a higher refractive index for the SiO_2 (for instance, $n_{Al_2O_3} = 1.73$ vs $n_{SiO_2} = 1.46$ at 1.95 eV) that offers better light confinement. Along with good solubility of rare-earth ions in such materials, their photonic application becomes conceivable. In fact, Er-doped hafnia-based waveguides or alumina-based waveguides have been developed for optical communications.

It is worth pointing out that in spite of well-developed approaches for the production of Si-NCs-SiO_2 and Si-NCs-Si_3N_4 composites, these methods are less addressed for the fabrication of Si-NCs embedded in other hosts. Only a few groups reported on Si-NCs-Al_2O_3 materials produced with ion implantation or electron beam evaporation (Núñez-Sánchez et al. 2009; Smit et al. 1986; Khomenkova et al. 2016). At the same time, in spite of the relative simplicity of the magnetron co-sputtering approach and its compatibility with CMOS technology, a deposition of Al_2O_3 remains to be challenged due to significant charging of the Al_2O_3 target, even with the RF sputtering approach. Only a few research groups reported successful formation of Si-NCs in the Al_2O_3 host produced by RF magnetron sputtering (Bi et al. 2006; Khomenkova et al. 2014, 2016).

The purpose of this session is to demonstrate the utility of magnetron sputtering for the fabrication of different dielectric materials with embedded Si-NCs and to compare the properties of these materials with those of Si-NCs-SiO_2 systems produced with the same routes.

6.4.1 Si-NCs-Al$_2$O$_3$

Materials produced by magnetron sputtering are rarely addressed (Baron et al. 2003; Korsunska et al. 2013; Khomenkova et al. 2016). However, the elaboration of such Si-doped Al$_2$O$_3$ thin films can give insight into the mechanism of the formation of ternary compounds and of Si-NCs in such hosts. Some recent results obtained for the layers grown by co-sputtering of Si and Al$_2$O$_3$ targets in argon plasma will be discussed. More details on the deposition procedure and annealing treatment can be found elsewhere (Korsunska et al. 2013; Khomenkova et al. 2014, 2016). It is worth pointing out that most publications consider composites with low Si to be excess. However, the samples with higher Si content demonstrate some unique optical and luminescent properties. Among two decomposition mechanisms, the phase separation via spinodal decomposition prevails (Khomenkova et al. 2016). The structural and luminescent properties of Si-rich-Al$_2$O$_3$ composites (as deposited and annealed at 1150°C) will be discussed hereafter as a function of Si content.

Raman scattering spectra showed the formation of amorphous Si clusters upon the deposition process for the films with Si excess higher than 30% (Figure 6.15a). Besides, the shift of the peak position of the transverse optic (TO) band to $\omega_{TO\text{-}a\text{-}Si} = 460 cm^{-1}$ was observed for Si-rich-A$_2$O$_3$ films, contrary to that detected for Si-rich-SiO$_2$ counterparts with similar Si excess ($\omega_{TO\text{-}a\text{-}Si} = 480 cm^{-1}$). This latter corresponds to the TO phonon peak position of relaxed amorphous silicon. The low-frequency shift observed for Si-rich-A$_2$O$_3$ samples is caused by the tensile stresses between the film and the fused quarts substrate due to mismatching in their lattice parameters.

Annealing treatment at $T_A = 1150°C$ results in the narrowing of TO phonon band and an increase in the intensity that is the evidence of Si-NCs formation. However, the amorphous Si phase is still present in the annealed films.

Grazing incidence XRD patterns of annealed Si-NCs-Al$_2$O$_3$ are shown in Figure 6.15c. Three main Si-related reflections corresponding to the (111), (220) and (311) family planes were detected. The annealing causes the appearance of Si-NCs with the sizes dependent on Si excess (Figure 6.15b). When it exceeds 50% ($x > 0.5$), the formation of Si-NCs with the mean size of about 14 nm occurs, as revealed by XRD data (Figure 6.15c), whereas similar samples submitted to rapid thermal annealing at 1050°C during 1 min showed the Si-NCs with mean sizes of about 5 nm. The comparison of these XRD data with those reported for Si-NCs-SiO$_2$ films with similar Si content show that the contribution of amorphous Si phase after conventional annealing is lower in Si-Al$_2$O$_3$ films (Khomenkova at al. 2016).

The observation of the same samples with *the HR-TEM method* shows an appearance of numerous Si-NCs, embedded in amorphous oxide hosts (Figure 6.15d). Contrast in the HR-TEM image arises from the coherent superposition of the primary and elastically scattered beams. For a thin enough imaged area, it is directly connected to the projected atomic structure of the Si-NCs. It is known that a crystal can be observed in HR-TEM with significant contrast if it is crystalline, in Bragg orientation with respect to the incident electrons and the thickness of the imaged area is not greater than two or three times the diameter of the Si-NCs (Schamm et al. 2008). Thus, amorphous and mis-oriented particles are excluded from this image. Moreover, the thin edges of the Si-NCs are too thin to be imaged as atomic columns. As a consequence, the size of the Si nanocrystals is underestimated. The minimum

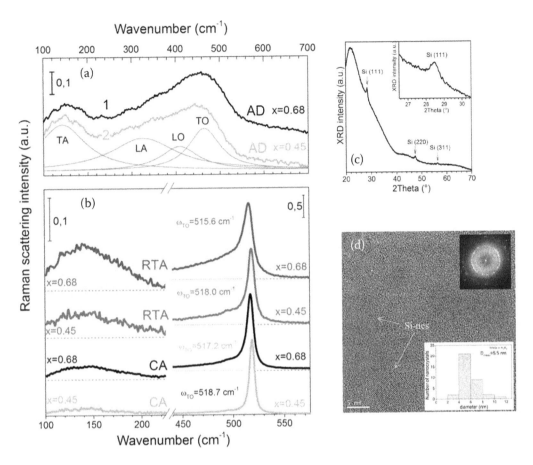

FIGURE 6.15 (a) Micro-Raman spectra of as-deposited, Si-rich Al_2O_3 films with $x=0.68$ (1) and $x=0.45$ (2). The deconvolution of curve 2 to four Si-phonon bands is also present. The spectra are offset for clarity. (b) Variation of micro-Raman spectra after rapid thermal annealing at 1050°C for 1 min and conventional annealing at 1150°C for 50 min on the same samples. (c) Grazing incidence XRD pattern from annealed $Si-Al_2O_3$. The inset shows the expanded presentation of the (111) Si peak; (d) High-resolution TEM image for annealed film on cross-sectional specimen. The insets show the associated Fast Fourier Transform of the image and the corresponding size distributions of Si-NCs. (From Khomenkova, L., Baran, M., Jedrzejewski, J., Bonafos,C., Paillard, V., Venger, Ye., Balberg, I. and Korsunska, N. 2016. Silicon nanocrystals embedded in oxide films grown by magnetron sputtering. *AIMS Materials Science* 3:538–561 and licensee under the terms of the Creative Commons Attribution License (http://creativecommons.org/licenses/by/4.0).)

detectable size that can be measured is equal to about 1.5 nm (i.e. 4–5 (110) planes). The average diameters have been extracted from the size distributions shown in the inset of Figure 6.15d. To build them, 35 nanocrystals have been measured for each sample (on 10 HR-TEM images). When the shape of the domain was not spherical (which is mostly the case), the major axis length has been considered. Touching nanocrystals that would lead to an overestimation of the mean size have been avoided in the statistics. The same has been applied for nanocrystals that are cut at the border of the images. The average diameter as measured by HR-TEM finally represents the size of monocrystalline domains. The Si nanoparticles could be larger and polycrystalline. For this reason, there is no sense in commenting on the shape of the size distribution. However, we can finally conclude that the Si nanocrystals are larger when Si nanoparticles are embedded in Al_2O_3, in comparison with those embedded in SiO_2. They represent the mean size of monocrystalline domains, i.e. Si nanoparticles could be significantly larger and polycrystalline. The number of Si-NCs was found to be higher in Si-Al_2O_3 than in Si-SiO_2, which is consistent with XRD data. The mean diameter of Si-NCs in Si-SiO_2 films is about 4 nm, whereas it exceeds 5.5 nm in Si-Al_2O_3 film. The discrepancy between the average size measured in the HREM images and the one obtained by XRD for the Si-Al_2O_3 sample also arises from the increased contribution of the largest nanocrystals in the XRD technique.

PL emission from as-deposited samples was not detected for $x > 0.5$, whereas for $x < 0.5$, only the peak at ~560 nm (2.21 eV) was observed. It was found to be similar to that detected in pure Al_2O_3 film and assigned to F_2^{2+} centres (Korsunska et al. 2013).

Annealing treatment at 1050°C–1150°C yields visible PL emission in a wider spectral range. Figure 6.16 represents room-temperature PL spectra of annealed samples. These spectra contain two broad PL bands, whose maxima are observed at 575–600 nm (2.06–2.15 eV) and 700–750 nm (1.65–1.77 eV), accompanied by a near-infrared tail (or weak band at 1.55–1.6 eV). These bands can be well-separated (for $x = 0.45$–0.5) or strongly overlapped. The first band consists of two components with maxima at ~2.06 eV and ~2.18 eV. The latter one is clearly seen in the sample with $x = 0.3$ being similar to PL emission from F_2^{2+} centres in Al_2O_3. At the same time, both emission components are strongly overlapped in the samples with $x > 0.3$ (Figure 6.16).

To elucidate the origin of PL emission from the films investigated, the PL spectra were measured also at 80 K, as described in (Khomenkova et al. 2016). However, for most samples, PL peak position was unaffected by the cooling. At the same time, few samples showed an increase in the intensity of the near-infrared tail. One can expect that this PL component belongs to the carrier recombination via Si-NCs. In spite of this result, one can conclude that the main contribution to the PL spectra in our samples is given by the carrier recombination through different defects located near the Si-NC/matrix interface (for instance, oxygen vacancies in Al_2O_3). On the other hand, the near-infrared PL band, which is assumed to be due to exciton recombination in Si-NCs, can be suppressed due to a high concentration of interface and matrix defects (in particular, the high intensity of the PL band at 700–750 nm). The behaviour of visible PL bands differs from that expected for quantum-confined Si-NCs, and in Si-NCs-Al_2O_3 samples. Therefore, PL can be ascribed to the interface and/or matrix defects. At the same time, the analysis of the PL spectrum shape allows ascribing the near-infrared PL component (780 to 900 nm) to the exciton recombination inside Si-NCs.

6.4.2 Si-NCs-HfO_2

The interest in Si-rich HfO_2 materials is caused, first of all, by their applications in microelectronics as "high-k" dielectrics alternative to SiO_2. One of the main applications of these materials is extended to floating gate non-volatile memory devices containing semiconductor nanoclusters embedded in the dielectric host. However, until the present time, the formation of Si-NCs in an HfO_2 matrix and in other high-k hosts is questionable. In spite of

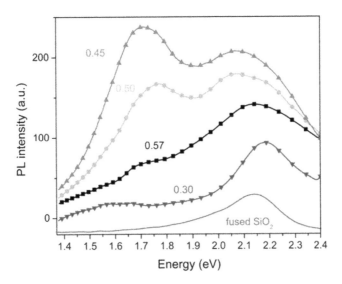

FIGURE 6.16 Room-temperature PL spectra of Si-rich-Al$_2$O$_3$ films. The values of the x are mentioned in the figures. Excitation wavelength is 488 nm.

the reported elaboration and fine characterisation of Si-rich-HfO$_2$ systems over a large composition range, any formation of Si nanoclusters was not revealed. It is worth pointing out that in most cases, the Si-rich HfO$_2$ composites are considered as those built from SiO$_2$ and HfO$_2$ units as (SiO$_2$)$_x$(HfO$_2$)$_{1-x}$. However, there are few reports of Si-rich HfO$_2$ (Khomenkova et al. 2013; An et al. 2013) where formation of ternary compound HfSiO$_x$ occurs as a random bonding of constituent elements. To investigate the formation of Si-NC in the HfO$_2$ host, composite thin films were grown on Si and fused quarts substrates by magnetron sputtering of the composed target. The Si content in the films was about 20 at % being distributed homogeneously in the HfO$_2$ bulk. The effect of annealing conditions on the structural and optical properties of the films was investigated. These results are compared with corresponding properties of Si-rich SiO$_2$ thin films with excess Si of about 20% grown in the same deposition unit and having the same thickness as Si-rich HfO$_2$ films.

Figure 6.17 shows the PL spectra of Si-rich-HfO$_2$ and Si-rich-SiO$_2$ films annealed at 950°C for 30 min in nitrogen flow. It should be noted in advance that the annealing of Si-rich-SiO$_2$ materials at 950°C does not result in the formation of crystallised Si-NCs. However, both types of materials show PL emissions of similar intensity in the red-near-infrared spectral range (Figure 6.17).

The comparison of pure SiO$_2$, pure HfO$_2$, Si-rich-SiO$_2$ and Si-rich HfO$_2$ films (Figure 6.17a) showed that this red emission can be due to carrier recombination inside Si clusters or in the vicinity of the cluster/matrix interface. The appearance of Si nanoclusters was observed by Raman scattering spectra at about 900°C, whereas a further increase of the temperature resulted in the variation of PL peak position and intensity (Figure 6.17b). The red emission was found to be very similar to that of Si-rich-SiO$_2$ materials (Figure 6.17a) and can thus be attributed to the formation of Si clusters in Si-rich-HfO$_2$ samples upon annealing treatment. The peak position of this PL band shifts to shorter wavelengths (from 820

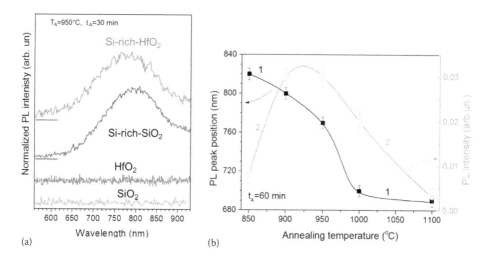

FIGURE 6.17 (a) PL spectra of pure SiO_2, pure HfO_2, Si-rich-SiO_2 and Si-rich HfO_2 films grown with the same deposition approach and annealed at 950°C for 30 min. (b) Variation of PL peak position (curve 1) and total PL intensity (curve 2) versus annealing temperature. Annealing time is 60 min. Excitation light wavelength is 488 nm.

nm to 690 nm) with T_A increase from 850°C to 1100°C at constant t_A (Figure 6.17b). This can be ascribed to the decrease of the mean size of Si clusters. The t_A increase at constant T_A results in the shift of PL peak position to longer wavelengths that can be caused by an increase of Si cluster sizes similar to the case observed for Si-rich SiO_2 films. The PL intensity shows a non-monotonous behaviour. The most efficient emission was detected for samples annealed at $T_A = 900°C–950°C$, $t_A = 60$ min. These samples were submitted to structural and chemical analysis by means of HR-TEM and atom-probe tomography (APT). The results obtained by both methods are presented in Figure 6.18.

The cross-sectional HR-TEM image of the annealed HfSiO$_x$ films revealed bright and dark contrasts that can be associated with Si-rich regions (bright) and Hf-rich ones (dark). This gives direct evidence of the Hf segregation process. The Hf-rich regions appear to be crystallised. In addition, the interfacial region between HfSiO film and Si substrate demonstrates the presence of two layers, namely a 2-nm thick SiO$_x$ layer followed by a 3-nm thick Hf-rich layer parallel to the Si substrate. A similar structure has been observed already in the case of ultrathin HfSiO layers and explained by the surface directed spinodal decomposition, taking into account the surface-to-volume ratio (Khomenkova et al. 2013).

The atom-probe tomography is a three-dimensional, high-resolution analytic microscopy, offering spatial mapping of atoms in materials. Nowadays, this technique is increasingly used for the study of semiconductor and dielectric materials. The APT technique is based on the field evaporation of surface atoms from tip-shaped specimens with a curvature radius lower than 50 nm. More details about this method can be found in Russel et al. (2011) and Talbot et al. (2012).

Atom-probe tomography allowed us to identify the microstructure of the HfSiO layer and to measure the composition of the different chemical zones and their spatial distributions. We have calculated iso-concentration surfaces for Hf and Si atoms. Figure 6.18 shows the surface

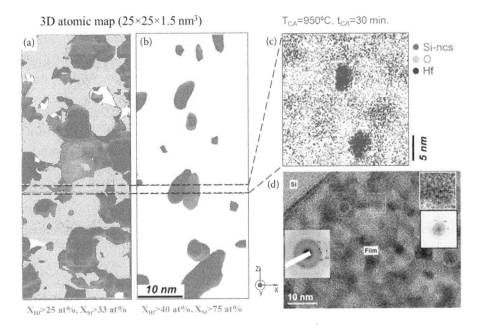

FIGURE 6.18   3D chemical iso-concentration surfaces of hafnium and silicon extracted from the 3D chemical maps. The thresholds on Hafnium and Silicon concentration are respectively (a) $X_{Hf} > 25$ at% and $X_{Si} > 33$ at% and (b) $X_{Hf} > 40$ at% and $X_{Si} > 75$ at%. (c) 3D atomic mapping in selected region (dash rectangle – volume 25 nm × 25 nm × 1.5 nm). For clarity, only silicon atoms belonging to pure Si-NCs are shown. (d) cross section HR-TEM micrograph of the same film. (Adapted with permission from *Mater Sci Eng B*, 177, Talbot et al., Atomic scale microstructures of high-k HfSiO thin films fabricated by magnetron sputtering, 717–720. Copyright 2012, Elsevier.)

for $X_{Hf} > 25$ at % (in the blue) and $X_{Si} > 33$ at % (in the red). These data provide the evidence of phase separation between SiO_2-rich and HfO_2-rich phases induced by the annealing treatment. Moreover, increasing the concentration threshold of iso-concentration (Figure 6.18b, where the threshold are $X_{Hf} > 40$ at% and $X_{Si} > 75$ at%) does not evidence Hf zones containing more than 40 at%, but clusters highly enriched in silicon. Figure 4c displays a top view in a selected region of the analysed volume, which contains the different chemical zones mentioned previously. Compositions of the three chemical regions were found to correspond to HfO_2, SiO_2 and pure Si. Thus, APT experiments revealed the presence of Si nanoclusters in the films after their high-temperature annealing. We should note that APT experiments can lead to a bias in the concentration measurement in the case of multi-phase sample analysis. This effect is due to the difference in evaporation fields between the different phases.

6.5 GE-NCS IN DIFFERENT DIELECTRIC HOSTS

The study of Si and Ge-based nanostructured materials is motivated by the prediction that quantum confinement of carriers leads to efficient luminescence despite the indirect nature of the energy gaps (Maeda et al. 1991; Maeda 1995; Ray et al. 2005; Sood et al. 2015; Wihl et al. 1972; Yamamoto et al. 1989; Zacharias and Fauchet 1997). Ge-NCs in the host matrix (SiO_2, Al_2O_3, HfO_2, etc.) layers can be produced by either ion implantation or by magnetron sputtering, in both cases followed by heat treatment (Das et al. 2012). The mechanisms involved in Ge nanocrystal growth are still controversial. The Ge-NCs have been found to exhibit visible luminescence at room temperature. However, the mechanism of visible luminescence of Ge nanocrystals is still disputed. Hereafter, the Ge-NCs' formation in different hosts will be considered and compared with that of Si-NCs.

The dark patches in all the figures are Ge nanocrystals. The Ge-NCs are almost spherical, with a diameter of 5–10 nm, being well dispersed in an amorphous Al_2O_3 matrix. The estimated size distribution of the nanocrystals for these samples was found to be approximately a Gaussian function with an average diameter of 7.1 nm. The change in Gibbs free energy formation of GeO (–111.8 kcal/mol) is much smaller than that of high-k Al_2O_3 (–378.2 kcal/mol) (Weast et al. 1990), which results in the oxidation of Al and the agglomeration of Ge atoms into nanocrystals in the Al_2O_3 matrix during thermal annealing at high temperatures.

Figure 6.19b shows a high-resolution TEM micrograph of Ge nanocrystals embedded in the HfO_2 matrix and annealed at 900°C, which exhibit clear lattice fringes. The average diameter of the nanocrystal is about 7.8 nm. The change in Gibbs free energy [ΔG] of formation (at 298.15 K) of GeO (–111.8 kcal/mol) is much smaller than that of high-k HfO_2 (–260.1 kcal/mol). Therefore, the change in Gibbs free energy is negative in the forward direction in the following reaction: $GeO_2 + Hf \rightarrow HfO_2 + Ge$. Hence, the mixture of HfO_2 and Ge has the lower Gibbs free energy in the co-sputtered film, resulting in the agglomeration of Ge atoms into nanocrystals.

Figure 6.19c shows the TEM image of Ge nanocrystals embedded in the SiO_2 matrix annealed at 900°C. The Gaussian fitting of the size distribution gives an average nanocrystal size of about 4.3 nm. The formation of Ge nanocrystals is attributed to the precipitation of Ge within the thermodynamically favourable SiO_2 layer during post-deposition annealing in N_2. The crystallisation process is a dynamic one, with nucleation and growth in addition to the migration of the Ge nanocrystals. It has been reported that the diffusion of Ge in SiO_2 and the nucleation of Ge depend on the annealing temperature. The size of the nanocrystals increases with increasing annealing temperatures due to the enhanced nucleation and growth process of Ge nanocrystals at the Si-SiO_2 interface. Furthermore,

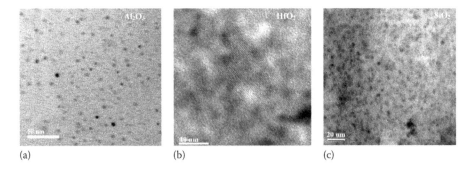

(a) (b) (c)

FIGURE 6.19 The plane-view TEM images of Ge-NCs embedded in Al_2O_3 (a), HfO_2 (b) and SiO_2 (c) matrix annealed at 900°C. (Adapted from Das et al. 2012 under the terms of the Creative Commons Attribution License (http://creativecommons.org/licenses/by/2.0).)

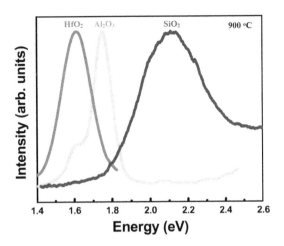

FIGURE 6.20   Room-temperature photoluminescence from Ge-NCs embedded in HfO_2, Al_2O_3 and SiO_2 hosts. Samples were annealed at 900°C. (Adapted from Das et al. 2012 under the terms of the Creative Commons Attribution License (http://creativecommons.org/licenses/by/2.0).)

a higher annealing temperature leads to an increase in the critical nucleus size and would also raise the barrier for nucleation.

Figure 6.20 shows the effect of a host matrix on room-temperature PL spectra of Ge-NCs embedded in SiO_2, Al_2O_3 and HfO_2 and annealed at a temperature of 900°C. The PL spectrum for Ge-SiO_2' samples indicates that the 2.11 eV peak originates due to radiative recombination in quantum-confined Ge nanocrystals. Two intense broad emission peaks are observed around 1.75 eV and 1.67 eV for samples Ge-Al_2O_3 and Ge-HfO_2, respectively.

The difference in PL peak energy between the samples may be attributed to the variation in average particle size in combination with the matrix-induced effect. In order to interpret the result quantitatively, a simple confinement model has been applied by considering electrons and holes confined independently in quantum dots of radius R

$$E_{nl} = E_g + \frac{\hbar^2}{2\mu_{e-h}}(\alpha_{nl}/R)^2 - 1.786e^2/kR,$$

where the second term represents the kinetic energy of electron and holes, and the last term denotes the Coulomb interaction term; μ_{e-h} is the reduced mass of excitons, k is the static dielectric constant (for Ge, $k = 16.3$), α_{nl} is the eigenvalue of the zeroth-order spherical Bessel function ($\alpha_{10} = \pi$) and E_g is the band gap energy of Ge = 0.66 eV. The calculated Ge-NCs' size, according to the quantum confinement model using this Equation, along with that estimated from the TEM micrograph, showed good agreement. For Ge-NCs embedded in HfO_2, Al_2O_3 and SiO_2 TEM data showed 7.8, 7.1 and 5.3 nm, while the PL data fitting with the confinement model give the next results: 7.1, 6.6 and 5.9 nm, correspondingly. There is a slight difference in the extracted size from the confinement theory and TEM observations. However, the carriers confined in the quantum dot in this case are under a finite potential, which has not been considered in the present confinement model.

Depending upon the host oxide matrix, the conduction and valence band offsets between the germanium nanocrystals and matrix are different, which leads to surrounding matrix-dependent confinement potential. The polarisation interface charge-induced nanocrystal band gap modification may also play an important role due to the difference in the dielectric constant of the host matrix and the nanocrystals. Therefore, these results give experimental evidence of the role of the dielectric constant and band offsets on the optical band gap of Ge-NCs bounded in different oxide matrices. Thus, one can expect that Ge-NCs will exhibit more of the promised properties in comparison with Si-NCs. Some of them were already discussed at the beginning of this chapter, whereas more detailed consideration is the subject of future publications.

In conclusion, some recent achievements in the fabrication of dielectric materials with embedded semiconductor nanoclusters and their characterisation by optical, electrical, structural and luminescent methods was demonstrated and discussed for Si-NCs and Ge-NCs embedded in different hosts. Among different dielectric materials, the main attention was paid to SiO_2, Al_2O_3 and HfO_2. The effect of the doping of these materials with Si and/or Ge on the formation of corresponding solid solutions, as well as their reaction on the thermal treatment, were described. The mechanisms of the formation of Si and/or Ge nanoclusters via the phase segregation process were discussed. The utility of magnetron sputtering for the fabrication of composite layers and superlattices was demonstrated on their basis.

REFERENCES

An, Y., Labbé, C., Khomenkova, L., Morales, M., Portier, X. and Gourbilleau, F. 2013. Microstructure and optical properties of Pr^{3+}-doped hafnium silicate films. *Nanoscale Res Lett* 8:43.

Amato, M., Palummo, M. Rurali, R. and Ossicini, S. 2014. Silicon-germanium nanowires: Chemistry and physics in play. *Chem Rev* 114:1371–1412.

Barbagiovanni, E. G., Lockwood, D. J., Simpson, P. J. and Goncharova, L. V. 2012. Quantum confinement in Si and Ge nanostructures. *J Appl Phys* 111:034307.

Baron, T., Fernandes, A., Damlencourt, J. F., De Salvo, B. and Martin, F. 2003. Growth of Si nanocrystals on alumina and integration in memory devices. *Appl Phys Lett* 82:4151–4153.

Barranco, A., Mejías, J. A., Espinós, J. P., Caballero, A., González-Elipe, A. R., and Yubero, F. 2001. Chemical stability of Si^{n+} species in SiO_x (x<2) thin films. *J Vac Sci Technol A* 19:136–144.

Bi, L. and Feng, J. Y. 2006. Nanocrystal and interface defects related photoluminescence in silicon-rich Al_2O_3 films. *J Lumin* 121:95–101.

Bulijan, M., Grenzer, J., Holý, V., Radić, N., Mišić-Radić, T., Levichev, S., Bernstorff, S., Pivac, B. and Capan, I. 2010. Structural and charge trapping properties of two bilayer $(Ge+SiO_2)/SiO_2$ films deposited on rippled substrate. *Appl Phys Lett* 97:163117.

Buuren, T. V., Dinh, L. N., Chase, L. L., Siekhaus, W. J. and Terminello, L. J. 1998. Changes in the electronic properties of Si nanocrystals as a function of particle size. *Phys Rev Lett* 80:3803.

Canham, L. T. 1986. Room temperature photoluminescence from etched silicon surfaces: The effects of chemical pretreatments and gaseous ambients. *J Phys Chem Sol* 47:363–373.

Canham, L. T. 1990. Silicon quantum wire array fabrication by electrochemical and chemical dissolution of wafers. *Appl Phys Lett* 57:1046–1048.

Canham, L. T. 1997. *Properties of Porous Silicon*. London: INSPEC.

Chan, M. Y., Lee P. S., Ho V. and Seng H. L.. 2007. Ge nanocrystals in lanthanide-based Lu_2O_3 high-k dielectric for nonvolatile memory applications. *J Appl Phys* 102:094307.

Changa, T.-C., Jian, F.-Y., Chen, S.-C. and Tsai, Y.-T. 2011. Developments in nanocrystal memory. *Mater Today* 14:608.

Chen, T. and Liu, Y. 2016. *Semiconductor Nanocrystals and Metal Nanoparticles.* Boca Raton, FL: CRC Press.

Cho, E. C., Green, M. A., Conibeer, G., Song, D., Cho, Y.-H., Scardera, G., Huang, S., Park, S., Hao, X. J., Huang, Y. and Dao, L. V. 2007. Silicon quantum dots in a dielectric matrix for all-silicon tandem solar cells. *Adv in OptoElectron*, Article ID 69578 (11 pages).

Colder, H., Marie, P. and Gourbilleau, F. 2008. The silicon-silicon oxide multilayers utilization as intrinsic layer on pin solar cells. *Thin Solid Films* 516:6930–6933.

Cosentino, S., Mirabella, S., Miritello, M., Nicotra, G., Lo Savio, R., Simone, F., Spinella, C. and Terrasi, A. 2011. The role of the surfaces in the photon absorption in Ge nanoclusters embedded in silica. *Nanoscale Res Lett* 6:135.

Cressler, J. D. 2016. *Silicon Earth: Introduction to Microelectronics and Nanotechnology*, Second Edition. Boca Raton, FL: CRC Press.

Das, K., NandaGoswami, M., Mahapatra, R., Kar, G. S., Dhar, D. and Acharya, H. N. 2004. Charge storage and photoluminescence characteristics of silicon oxide embedded Ge nanocrystal trilayer structures. *Appl Phys Lett* 84:1386–1388.

Das, S., Das, K., Singha, R. K., Dhar, A. and Ray, S. K. 2007. Improved charge injection characteristics of Ge nanocrystals embedded in hafnium oxide for floating gate devices. *Appl Phys Lett*, 91:233118–233120.

Das, S., Aluguri R., Manna S., Singha, R., Dhar, A., Pavesi, L. and Ray, S. K. 2012. Optical and electrical properties of undoped and doped Ge nanocrystals. *Nanoscale Res Lett* 7:143.

Dehan, E., Temple-Boyer, P., Henda, R., Pedroviejo, J. J. and Scheid, E. 1995. Optical and structural properties of SiO_x and SiN_x materials. *Thin Solid Films* 266:14–19.

Delerue, C., Allan, G. and Lannoo, M. 1993. Theoretical aspects of the luminescence of porous silicon. *Phys Rev B* 48:11024.

Di, D., Perez-Wurfl, I., Conibeer, G. and Green, M. A. 2010. Formation and photoluminescence of Si quantum dots in SiO_2/Si_3N_4 hybrid matrix for all-Si tandem solar cells. *Sol Energy Mater Sol Cells* 94:2238–2243.

Ding, L., Yu, M. B., Xiaoguang T., Lo, G. Q., Tripathy, S. and Chen, T. P. 2011. Laterally-current-injected light-emitting diodes based on nanocrystalline-Si/SiO_2 superlattice. *Opt Express* 19:2729–2738.

Doğan, I. and van de Sanden, M. C. M. 2013. Direct characterization of nanocrystal size distribution using Raman spectroscopy. *J Appl Phys* 114:134310.

Fauchet, P. M. and Campbell, I. H. 1988. Raman spectroscopy of low-dimensional semiconductors. *Crit Rev Solid State Mater Sci.* 14(S1):s79–s101.

Fu, S. W., Chen, H. J., Wu, H. T., Chen S. P. and Shih, C. F. 2016. Enhancing the electroluminescence efficiency of Si NC/SiO_2 superlattice-based light-emitting diodes through hydrogen ion beam treatment. *Nanoscale* 8:7155–7162.

Gourbilleau, F., Khomenkova, L., Bréard, D., Dufour, C. and Rizk, R. 2009a. Rare Earth (Er, Nd) doped Si nanostructures for integrated photonics. *Physica E* 41:1034.

Gourbilleau, F., Dufour, C., Rezgui, B. and Brémond, G. 2009b. Silicon nanostructures for solar cell applications. *Mater Sci Eng B* 159–160:70–73.

Gourbilleau, F., Ternon, C., Maestre, D., Palais, O. and Dufour, C. 2009c. Silicon-rich SiO_2/SiO_2 multilayers: A promising material for the third generation of solar cell. *J Appl Phys* 106:013501.

Grachev, D. A., Legkov, A. M., Chunin, I. I. and Ershov, A. V. 2017. Luminescent property modification of SiO_x/Al_2O_3 multilayers by annealing and hydrogenation. *J Phys Conf Ser* 816:012007.

Haas, S., Schneider, F., Himcinschi, C., Klemm, V., Schreiber, G., von Borany, J. and Heitmann, J. 2013. Ge nanoparticle formation by thermal treatment of rf-sputtered $ZrO_2/ZrGe_2O_3$ superlattices. *J Appl Phys* 113:044303.

Hayashi, S., Ito, M. and Kanamori, H. 1982. Raman study of gas-evaporated germanium microcrystals. *Solid State Commun* 44:75–79.

Hessel, C. M., Reid, D., Panthani, M. G., Rasch, M. R., Goodfellow, B. W., Wei, J., Fuji, H., Akhavan, V. and Korgel, B. A. 2011. Synthesis of ligand-stabilized silicon nanocrystals with size-dependent photoluminescence spanning visible to near-infrared wavelengths. *Chem Mater* 24:393. https://doi.org/10.1021/cm2032866.

Hijazi, K., Khomenkova, L., Cardin, J., Gourbilleau, F. and Rizk, R. 2009. Structural and optical characteristics of Er-doped SRSO layers deposited by the confocal sputtering technique. *Physica E* 41:1067–1070.

Hubner, K. 1980. Chemical bond and related properties of SiO_2. VII. Structure and electronic properties of the SiO_x region of Si-SiO_2 interfaces. *Phys Stat Sol A* 61:665–673.

Ilse, K., Schneider, T., Ziegler, J., Sprafke, A. and Wehrspohn, R. B. 2016. Integrated low-temperature process for the fabrication of amorphous Si nanoparticles embedded in Al_2O_3 for non-volatile memory application. *Phys Stat Sol A* 213:2446–2451.

Innocenzi, P. 2003. Infrared spectroscopy of sol-gel derived silica-based films: a spectra-microstructure overview. *J Non-Cryst Sol* 316:309.

Ischenko, A. A., Fetisov, G. V. and Aslanov, L. A. 2015. *Nanosilicon: Properties, Synthesis, Applications, Methods of Analysis and Control.* Boca Raton, FL: CRC Press.

Kamata, Y. 2008. High-k/Ge MOSFETs for future nanoelectronics. *Mater Today* 11:30–38.

Kanemitsu, Y., Masumoto, H., Uto, Y. and Maeda, Y. 1992. On the origin of visible photoluminescence in nanometer-size Ge crystallites. *Appl Phys Lett* 61:2187–2189.

Kartopu, G., Sapelkin, A. V., Karavanskii, V. A., Serincan, U. and Turan, R. 2008. Structural and optical properties of porous nanocrystalline Ge. *J Appl Phys* 103:113518.

Khomenkova, L., Baran, M., Jedrzejewski, J., Bonafos, C., Paillard, V., Venger, Ye., Balberg, I. and Korsunska, N. 2016. Silicon nanocrystals embedded in oxide films grown by magnetron sputtering. *AIMS Mater Sci* 3:538–561.

Khomenkova, L., Baran, M., Kolomys, O., Strelchuk, V., Kuchuk, A. V., Kladko, V. P., Jedrzejewski, J., Balberg, I., Goldstein, Y., Marie, P., Gourbilleau, F. and Korsunska, N. 2014 Comparative investigation of structural and optical properties of Si-rich oxide films fabricated by magnetron sputtering. *Adv Mat Res* 854:117–124.

Khomenkova, L., Gourbilleau, F., Cardin, J., Jambois, O., Garrido, B. and Rizk, R. 2009. Long lifetime and efficient emission from Er^{3+} ions coupled to Si nanoclusters in Si-rich SiO_2 layers. *Journal of Luminescence* 129:1519–1523.

Khomenkova, L., Korsunska, N., Yukhimchuk, V., Jumayev, B., Torchynska, T., Vivas Hernandez, A., Many, A., Goldstein, Y., Savir, E. and Jedrzejewski, J. 2003. Nature of visible luminescence and its excitation in Si-SiO_x systems. *J Lumin* 102:705–711.

Khomenkova, L., Labbé, C., Portier, X., Carrada, M. and Gourbilleau, F. 2013. Undoped and Nd^{3+} doped Si-based single layers and superlattices for photonic applications. *Phys Stat Sol A* 210:1532–1543.

Khomenkova, L., Portier, X., Cardin, J. and Gourbilleau, F. 2010. Thermal stability of high-*k* Si-rich HfO_2 layers grown by RF magnetron sputtering. *Nanotechnology* 21:285707.

Khomenkova, L., Sahu, B. S., Slaoui, A. and Gourbilleau, F. 2011. Hf-based high-k materials for Si nanocrystal floating gate memories. *Nanoscale Res Lett* 6:172.

Khriachtchev, L., 2009. *Silicon Nanophotonics: Basic Principles, Present Status, and Perspectives* Singapore: Pan Stanford.

Khriachtchev, L. 2016. *Silicon Nanophotonics: Basic Principles, Present Status, and Perspectives,* 2nd Edition. New York: Pan Stanford.

Kirk, C. T. 1988. Quantitative analysis of the effect of disorder-induced mode coupling on infrared absorption in silica. *Phys Rev B* 38:1255.

Korsunska, N., Khomenkova, L., Kolomys, O., Strelchuk, V., Juchuk, A., Kladko, V., Stara, T., Oberemok, O., Romanyuk, B., Marie, P., Jedrzejewski, J. and Balberg, I. 2013. Si-rich Al_2O_3 films grown by RF magnetron sputtering: Structural and photoluminescence properties versus annealing treatment. *Nanoscale Res Lett* 8:273.

Koshida, N. 2009. *Device Applications of Silicon Nanocrystals and Nanostructures.* New York: Springer Science–Business Media.

Koshizaki, N., Umehara, H. and Oyama, T. 1998. XPS characterization and optical properties of Si/ SiO_2, Si/Al_2O_3 and Si/MgO co-sputtered films. *Thin Solid Films* 325:130–136.

Kumar, V. 2007. *Nanosilicon.* Oxford: Elsevier.

Lange, P. 1989. Evidence for disorder-induced vibrational mode coupling in thin amorphous SiO_2 films. *J Appl Phys* 66:201.

Ledoux, G., Guillois, O., Porterat, D., Reynaud, C., Huisken, F., Kohn, B. and Paillard, V. 2000. Photoluminescence properties of silicon nanocrystals as a function of their size. *Phys Rev B* 62:15942.

Lehmann, V. and Gösele, U. 1991. Porous silicon formation: A quantum wire effect. *Appl Phys Lett* 58:856–858. https://doi.org/10.1063/1.104512.

Lehninger, D., Beyer, J. and Heitmann, J. 2018. A review on Ge nanocrystals embedded in SiO_2 and high-k dielectrics. *Phys Stat Sol A* 155:1701028.

Li, Y., Liang, P., Hu, Z., Guo, S., You, Q., Sun, J., Xu, N. and Wu, J. 2014. Enhancement and stability of photoluminescence from Si nanocrystals embedded in a SiO_2 matrix by H_2-passivation. *Appl Surf Sci* 300:178–183.

Liang, C.-H., Debieu, O., An, Y.-T., Khomenkova, L., Cardin, J. and Gourbilleau, F. 2012a. Effect of the Si excess on the structure and the optical properties of Nd-doped Si-rich silicon oxide. *Journal of Luminescence* 132:3118–3121.

Liang, C.-H., Cardin, J., Khomenkova, L. and Gourbilleau, F. 2012b. Effect of annealing treatment on Nd-SiOx thin film properties. *Proc. SPIE* 8431:84311Y.

Lisovskii, I. P., Litovchenko, V. G., Lozinskii, V. B., Frolov, S. I., Flietner, H., Fusel, W. and Schmidt, E. G. 1995. IR study of short-range and local order in SiO_2 and SiO_x films. *J Non-Cryst Sol* 187:91–95.

Maeda, Y., Tsukamoto, N., Yazawa, Y., Kanemitsu, Y. and Masumoto, Y. 1991. Visible photoluminescence of Ge microcrystals embedded in SiO_2 glassy matrices. *Appl Phys Lett* 59:3168–3170.

Maeda, Y. 1995. Visible photoluminescence from nanocrystallite Ge embedded in a glassy SiO_2 matrix: Evidence in support of the quantum-confinement mechanism. *Phys Rev B* 51:1658–1670.

Mastronardi, M. L., Maier-Flaig, F., Faulkner, D., Henderson, E. J., Kübel, C., Lemmer, U. and Ozin, G. A. 2012. Size-dependent absolute quantum yields for size-separated colloidally-stable silicon nanocrystals. *Nano Lett* 12:337.

Min, K. S., Shcheglov, K. V., Yang, C. M. and Atwater, H. A. 1996. The role of quantum-confined excitons vs defects in the visible luminescence of SiO_2 films containing Ge nanocrystals. *Appl Phys Lett* 68:2511–2513.

Nesbit, L. A. 1985. Annealing characteristics of Si-rich SiO_2 films. *Appl Phys Lett* 46:38–40.

Ni, Z., Pi, X. D. and Yang, D. 2014. Doping Si nanocrystals embedded in SiO_2 with P in the framework of density functional theory. *Phys Rev B* 89:035312.

Núñez-Sánchez, S., Serna, R., García López, J., Petford-Long, A. K., Tanase, M. and Kabius, B. 2009: Tuning the Er^{3+} sensitization by Si nanoparticles in nanostructured as-grown Al_2O_3 films. *J Appl Phys* 105:013118.

Oda, S. and Ferry, D. 2005. *Silicon Nanoelectronics.* Boca Raton, FL: CRC Press.

Olsen, J. E. and Shimura, F. 1989. Infrared reflection spectroscopy of the SiO_2-silicon interface. *J Appl Phys* 66:1353.

Ono, H., Ikarashi, T., Ando, K. and Kitano, T. 1998. Infrared studies of transition layers at SiO_2/Si interface. *J Appl Phys* 84:6064.

Ossicini, L., Pavesi, L. and Priolo, F. 2003. *Light Emitting Silicon for Microphotonics.* Berlin: Springer-Verlag.

Pai, P. G., Chao, S. S., Takagi, Y. and Lukovsky, G. 1986. Infrared spectroscopic study of SiO_x films produced by plasma enhanced chemical vapor deposition. *J Vac Sci Technol* A 4:689.

Pavesi, L., Gaponenko, S. V. and Dal Negro, L. 2003. Towards the First Silicon Laser. Dordrecht, The Netherlands: NATO Science Series, Kluwer Academic Publishers.

Pavesi, L. and Turan, R. 2010. *Silicon Nanocrystals: Fundamentals, Synthesis and Applications.* Weinheim, Germany: Wiley-VCH Verlag GmbH & Co. KGaA.

Pratibha Nalini, R., Dufour, C., Cardin, J. and Gourbilleau, F. 2011. New silicon based multilayers for solar cell applications. *Nanoscale Res Lett* 6:156–161.

Pratibha Nalini, R., Khomenkova, L., Debieu, O., Cardin, J., Dufour, C., Carrada, M. and Gourbilleau, F. 2012. SiO_x/SiN_y multilayers for photovoltaic and photonic applications. *Nanoscale Res Lett* 7:124.

Pratibha Nalini, R., Marie, P., Cardin, J., Dufour, C., Dimitrakis, P., Normand, P., Carrada, M. and Gourbilleau, F. 2011. Enhancing the optical and electrical properties of Si-based nanostructured materials. *Energy Procedia* 10:161–166.

Ray, S. K., Maikap, S., Banerjee, W. and Das, S. 2013. Nanocrystals for silicon-based light-emitting and memory devices. *J Phys D Appl Phys* 46:153001.

Ray, S. K. and Das, K. 2005. Luminescence characteristics of Ge nanocrystals embedded in SiO_2 matrix. *Opt Mater* 27:948–952.

Rochet, F., Dufour, G., Roulet, H., Pelloie, B., Perrière, J., Fogarassy, E., Slaoui, A. and Froment, M. 1988. Modification of SiO through room-temperature plasma treatments, rapid thermal annealing, and laser irradiation in a nonoxidizing atmosphere. *Phys Rev B* 37:6468–6477.

Roussel, M., Talbot, E., Pratibha Nalini, R., Gourbilleau, F. and Pareige, P. 2011. Phase transformation in SiO_x/SiO_2 multilayers for optoelectronics and microelectronics applications. *Ultramicroscopy* 132:290–294.

Schmidt, J. U. and Schmidt, B. 2003. Investigation of Si nanocluster formation in sputter deposited silicon sub-oxides for nanocluster memory structures. *Mat Sci Eng B* 101:28–33.

Schamm, S., Bonafos, C., Coffin, H., Cherkashin, N., Carrada, M., Ben Assayag, G., Claverie, A., Tencé, M. and Colliex, C. 2008. Imaging Si nanoparticles embedded in SiO_2 layers by (S)TEM-EELS. *Ultramicroscopy* 108:346–357.

Siffert, P. and Krimmel, E. F. 2004. Silicon: Evolution and Future of a Technology. New York: Springer.

Smit, M. K., Acket, G. A. and van der Laan, C. J. 1986. Al_2O_3 films for integrated optics. *Thin Solid Films* 138:171–181.

Sohrabi, F., Nikniazi, A. and Movla, H. 2013. Optimization of Third generation nanostructured silicon-based solar cells. In *Solar Cells*, A. Morales-Acevedo (Ed.), chapter 1, p. 1–25, Rijeka: InTech.

Sood, A. K., Zeller, J. W., Richwine, R. A., Puri, Y. R., Efstathiadis, H., Haldar, P., Dhar, N. K. and Polla, D. L. 2015. SiGe based visible-NIR photodetector technology for optoelectronic applications, advances in optical fiber technology. In *Fundamental Optical Phenomena and Applications*, Dr Moh. Yasin (Ed.), Rijeka: InTech. Available from: https://www.intechopen.com/books/advances-in-optical-fiber-technology-fundamental-optical-phenomena-and-applications/sige-based-visible-nir-photodetector-technology-for-optoelectronic-applications.

Steimle, R. F., Muralidhar, R., Rao, R., Sadd, M., Swift, C. T., Yater, J., Hradsky, B., Straub, S., Gasquet, H., Vishnubhotla, L., Prinz, E. J., Merchant, T., Acred, B., Chang, K. and White, B. E. Jr. 2007. Silicon nanocrystal non-volatile memory for embedded memory scaling. *Microelectron Reliability* 47:585–592.

Talbot, E., Roussel, M., Khomenkova, L., Gourbilleau, F. and Pareige, P. 2012. Atomic scale microstructures of high-k HfSiO thin films fabricated by magnetron sputtering. *Mater Sci Eng B* 177:717–720.

Ternon, C., Gourbilleau, F., Portier, X., Voivenel, P. and Dufour, C. 2002. An original approach for the fabrication of Si/SiO_2 multilayers using reactive magnetron sputtering. *Thin Solid Films* 419:5–10.

Torchynska, T. V. and Vorobiev, Yu. V. 2010. *Nanocrystals and Quantum Dots of Group IV Semiconductors.* Stevenson Ranch, CA: American Scientific Publishers.

Tsu, D. V., Lucovsky, G. and Davidson, B. N. 1989. Effects of the nearest neighbors and the alloy matrix on SiH stretching vibrations in the amorphous SiOr:H (0<r<2) alloy system. *Phys. Rev. B* 40:1795.

Tucker, M. G., Keen, D. A., Dove, M. T. and Trachenko, K. 2005. Refinement of the Si–O–Si bond angle distribution in vitreous silica. *J Phys Cond Matt* 17:S67–S76.

Vaughn II, D. D. and Schaak, R. E. 2013. Synthesis, properties and applications of colloidal germanium and germanium-based nanomaterials. *Chem Soc Rev* 42:2861.

Weast, R. C., Lide, D. R., Astle, M. J. and Beyer, W. H. (Eds). 1990. *CRC Handbook of Chemistry and Physics: A Ready-Reference Book of Chemical and Physical Data.* 70th Edition. Boca Raton: CRC.

Wihl, M., Cardona, M. and Tauc, J. 1972. Raman scattering in amorphous Ge and III-V compounds. *J Non-Cryst Sol* 8–10:172–178.

Yamamoto, K., Hayashi, S. and Fujii, M. 1989. Quantum size effects in Ge microcrystals embedded in SiO_2 thin films. *Jpn J Appl Phys* 28:L1464–L1466.

Zacharias, M. and Fauchet, P. M. 1997. Blue luminescence in films containing Ge and GeO_2 nanocrystals: The role of defects. *Appl Phys Lett* 71:380–382.

Zacharias, M., Heitmann, J., Scholz, R. and Kahler, U. 2002. Size-controlled highly luminescent silicon nanocrystals: A SiO/SiO_2 superlattice approach. *Appl Phys Lett* 80:661–663.

Zheng, T. and Li, Z. 2005. The present status of Si/ SiO_2 superlattice research into optoelectronic applications. *Superlattices Microstruct* 37:227–247.

Appendix

TABLE AI.1   Abbreviation, Chemical Name and Molecular Formula of Some of the Chemical Compounds Referred to in Chapters 3 and 4

Abbreviation	Chemical Name	Molecular Formula
PMMA	Poly(methyl methacrylate)	www.sigmaaldrich.com
PBMA	Poly(butyl methacrylate)	www.sigmaaldrich.com
PANI	Poly(aniline)	$(C_6H_7N)_x$ www.sigmaaldrich.com
PVA	Poly(vinyl alcohol)	$[-CH_2CHOH-]_n$

(*Continued*)

TABLE AI.1 (CONTINUED)   Abbreviation, Chemical Name and Molecular Formula of Some of the
Chemical Compounds Referred to in Chapters 3 and 4

Abbreviation	Chemical Name	Molecular Formula
PTSC	Pyridinecarboxaldehyde thiosemicarbazone	$C_7H_8N_4S$ www.sigmaaldrich.com
PAA	Poly(acrylic acid)	$(C_3H_4O_2)_n$ http://pslc.ws/macrog/acrylate.htm
PVI	Poly(vinyl imidazole)	$C_{15}H_{18}N_6X_2$
PVP	Poly(vinyl pyrrolidone)	$(C_6H_9NO)_n$
P(AAm-co-IA)	Poly(acrylamide-co-itaconic acid)	—
PPy	Poly(pyrrole)	$(C_4H_2NH)n$
TEMED	Tetramethylethylenediamine	$(CH_3)_2NCH_2CH_2N(CH_3)_2$
HDPE	High-density polyethylene	$(C_2H_4)_n$

(*Continued*)

TABLE AI.1 (CONTINUED)   Abbreviation, Chemical Name and Molecular Formula of Some of the
Chemical Compounds Referred to in Chapters 3 and 4

Abbreviation	Chemical Name	Molecular Formula
TGA	Thioglycolic acid	 $HSCH_2CO_2H$
PEG	Polyethylene glycol	 $C_{2n}H_{4n+2}O_{n+1}$
PDDA	Polydiallyl dimethyl ammonium chloride	$C_8H_{16}ClN$

https://www.sigmaaldrich.com/, http://pslc.ws/macrog/acrylate.htm.

Index